T0299567

Praise for *Our Future is Biotech*

"Andrew presents an invigorating vision of biotech's profound impact on our lives, underpinned by a rich tapestry of scientific innovation and human progress. *Our Future is Biotech* is a clarion call to those mired in pessimism, offering an informed, optimistic outlook grounded in tangible advancements and potential. With its accessible narrative and compelling evidence, it not only educates but inspires readers to engage with the biotech revolution – both for personal empowerment and as savvy investors. As the book reveals the sector's capacity to reshape industries from healthcare to clean energy, it's an essential read for anyone committed to understanding the technologies that will define our future."

—Christian Angermayer, Biotech Entrepreneur, Founder of Apeiron Investment Group

"A tour-de-force and a book that I found unputdownable. Andrew has seen the way for biotech and brings a unique, insightful and typically plain-English perspective to us all. This is a must-read and highly recommended."

—Barry Smith, Founder of Sunday Times' Fast Track 100 probiotics company, Symprove

"An excellent analysis of the life-changing scientific developments occurring in the world right now and a good prod in the ribs to governments and regulators to improve the funding environment to ensure we all benefit!"

—Jim Wilkinson, CFO of Oxford Science Enterprises

"*Our Future Biotech* serves as an essential guide through the transformative world of health technology. This book brilliantly maps out the intricate challenges and opportunities of the biotech sector in an era of extended life expectancy, increasing burden of chronic diseases, and widening health inequalities. With its thorough analysis and forward-looking solutions concerning new technologies, financial strategies, and healthcare reforms, it empowers readers to not only understand but also contribute to the evolution of healthcare. The inspiring 'The Future is Bright' chapter left me optimistic about our ability to harness innovation for better health outcomes. A must-read for anyone interested in the future of human health."

—Jo Pisani, Chair of the Precision Health Technology Accelerator and several European biotech companies, former head of PwC UK Pharma and Life Sciences

"Throughout our history, nature has sustained and nurtured humanity. In our complex modern world, more than ever before, we need to look again to nature to inspire and unlock innovative solutions and new technologies that can transform our way of life and protect and sustain the environment and biological support systems of our precious planet. As Andrew Craig so succinctly states in the title of his timely and erudite book: *Our Future is Biotech*. As a 'non-scientist', Andrew's grasp of the key concepts and complex fundamentals that underpin the application of biotechnology to solve problems in such diverse areas as human health, sustainable agriculture, waste management and clean energy production is extraordinary. He delivers a clear, comprehensive, accessible and unbiased view of the industry and its potential that is easily accessible to a wide readership. This is a must read for the shrewd investor with an eye to the significant opportunities emerging from the sustainable and life-changing new technologies emerging from this industry, but also for those more broadly interested in a guide to understanding biotechnology and its importance to the future of humanity and our planet."

—Dr Victoria Gordon, co-Founder of QBiotics Group, former Chairman of the Australian Rainforest Foundation

"In a world of ever-greater complexity, biotechnology is at the forefront of the complex: from the medicines we take, to the medical devices that are implanted, to the myriad of apps directing our diet and exercise regimes. Every day we are impacted by the products of this 'biotech revolution'. Even those who specialise in biotechnology can struggle to keep up! *Our Future is Biotech* is an outstanding, easy to read, anatomy of the impacts that biotechnology will have on all of us, what we need to understand and what we might need to do about it. I urge anyone with an interest in the future to read it!"

—Dr Eliot Forster, Professor of Molecular and Clinical Cancer Medicine (University of Liverpool), CEO of Levicept, Chairman of Avacta plc

"This is an important and extremely timely book which covers many key aspects of the biotech industry. Andrew provides excellent context for all readers no matter what level of knowledge and dives with appropriate depth into so many important areas of dramatic innovation. I thoroughly recommend this book as a clear guide for anyone interested in the sector and its impact on all of us, now and in the future."

—David Browning, biotech industry veteran, former Chairman of Oxford Bioscience Network (OBN)

"At bd-capital we look to invest in companies that are on trend with a unique point of difference. Biotech is a fertile hunting ground for our company, because there are large and growing unmet human needs, particularly as we all live longer. Andrew Craig's book places a good deal of emphasis on the crucial and poorly understood role of the microbiome in human health and in the inexorable rise of so many diseases of modernity, including IBS, IBD and a good deal else, which the evidence suggests may have much to do with a compromised microbiome. This is highly aligned with our stance and the reason our first portfolio investment was in the probiotics business, Symprove – a company Andrew holds in high regard. Ethical investing is at its best when good returns on capital, can be paired with making the world a better place. This is how we feel about our investment in Symprove and about the biotech sector more generally."

—Richard Baker, Managing Partner of bd-capital, former Chairman of Whitbread plc, CEO of Alliance Boots plc

"*Our Future is Biotech* is one of the most positive accounts of the potential of biotechnology that I have read. It emphasises the important interaction between adequate funding for research and the potential benefits to human health, making the case that if there is adequate funding then nothing is beyond our abilities to improve our existence on planet earth."

—Professor Ingvar Bjarnason MD, MSc, FRCPath, FRCP(Glasg), DSc, Professor of Digestive Diseases, Consultant Physician, Gastroenterologist

"This is a must-read. In a world of volatility and short-term thinking, Andy looks at the longer term trends that matter and identifies how we should be taking advantage of them. He has a lovely way of writing – inciteful yet humble, knowledgeable yet open to new and fresh ideas and opportunities. I recommend *Our Future is Biotech* to anyone that wants to know more about the future and how we can all benefit from the greatest period of technological and scientific change in human history."

—Mike Seabrook, veteran UK stockbroker, Head of Oberon Capital

"Andrew Craig has shone a bright spotlight on the extraordinary opportunities that will be created by 'biotech' in the coming decades, how this is driven by the convergence and combination of numerous life science disciplines, and how it will ultimately provide the answers to most of the world's problems! *Our Future is Biotech* is an excellent and positive review of the exponential progress made in innovative life sciences over the past decades and the life changing potential it holds – on a timescale that's meaningful for all of us. A thorough, insightful and uplifting read."

— Alastair Smith, Chief Executive of Avacta Group plc

"Andrew Craig's book is clear and direct. This is about why and how biotech is transforming health and delivering massive social, environmental and economic value – not just in impacting the major diseases that ruin and end lives prematurely such as diabetes, Alzheimer's and cancers, but also how biotech provides a key to developing more sustainable agriculture and environment. I was particularly struck by the section 'Biotech Miracles' that explains how biotech is providing the tools – drugs, diagnostics, devices and data/AI – to cure many diseases previously thought uncurable, to give hope in the fight against emerging health 'tsunamis' such as microbial resistance, and to extend old age and health in old age. *Our Future is Biotech* is about the cures, understanding, medicines, diagnostics and devices that allow us to live healthier, happier and longer lives in parallel with saving trillions of health dollars. It is a very cogent and well-argued case."

—Hugo Tewson, co-Founder of Chronic Foundation, Chairman of Digostics Limited

Our Future is Biotech

A Plain English Guide to How a Tech Revolution is Changing Our Lives and Our Health for the Better

ANDREW CRAIG

First published by John Murray Business in 2024
An imprint of John Murray Press

2

Copyright © Plain English Finance Ltd 2024

The right of Andrew Craig to be identified as the Author of the Work has been asserted by him in accordance with the Copyright, Designs and Patents Act 1988.

All rights reserved. No part of this publication may be reproduced, stored in a retrieval system, or transmitted, in any form or by any means without the prior written permission of the publisher, nor be otherwise circulated in any form of binding or cover other than that in which it is published and without a similar condition being imposed on the subsequent purchaser.

This book is for information or educational purposes only and is not intended to act as a substitute for medical advice or treatment. Any person with a condition requiring medical attention should consult a qualified medical practitioner or suitable therapist.

The author and publisher wish to make clear that the information contained in this book, including the opinions expressed, is offered as guidance only and is not intended to take the place of advice from a professional financial advisor. Any application of the investment principles contained in this book is at the reader's sole discretion and risk and the author and publisher assume no responsibility for any losses that may ensue.

References to companies and/or interests in this book are for illustrative purposes only. Nothing in this book constitutes a solicitation, recommendation, endorsement or offer by the author or publisher to deal in, buy, sell or hold any securities or other financial instruments in any of the companies and/or interests referenced in this book.

From time to time the author and certain investment vehicles managed by the author may deal in, buy, sell or hold investments in some of the companies and/or interests referenced in this book.

A CIP catalogue record for this title is available from the British Library

Hardback ISBN 978-1-39980-0-174
Trade Paperback ISBN 978-1-39980-0-198
UK ebook ISBN 978-1-39980-0-204
US ebook ISBN 978-1-39980-0-358

Typeset by KnowledgeWorks Global Ltd.

Printed and bound in Great Britain by Clays Ltd, Elcograf S.p.A.

John Murray Press policy is to use papers that are natural, renewable and recyclable products and made from wood grown in sustainable forests. The logging and manufacturing processes are expected to conform to the environmental regulations of the country of origin.

John Murray Press
Carmelite House
50 Victoria Embankment
London EC4Y 0DZ

John Murray Business
123 S. Broad St., Ste 2750
Philadelphia, PA 19109

John Murray Press, part of Hodder & Stoughton Limited

An Hachette UK company

Contents

This book is dedicated to Neville Craig, to David and Christine Hardy, and in loving memory of Mrs Gillian Craig (1949–2017).

About the author

Andrew Craig is a best-selling finance author, entrepreneur, and the founder of personal finance website www.plainenglishfinance.com. His stated mission with Plain English Finance is 'to improve the financial affairs of as many people as possible'.

Andrew's first book, *How to Own the World*, has been one of the top-selling personal finance books in the UK for the last several years, and currently enjoys more than 4,000 reviews across Amazon, Audible and Goodreads.

Since founding Plain English Finance in 2011, Andrew has appeared in numerous national and specialist financial publications including: *The Mail on Sunday*, *The Financial Times*, *The Mirror*, *City A.M.*, *The Spectator*, *Shares* and *MoneyWeek* magazines, *YourMoney*, *This is Money* and *Money Observer*. He has been interviewed on Sky Television, Bloomberg and Shares Radio and on IGTV, was featured in Michael Winterbottom's 2015 documentary-comedy *The Emperor's New Clothes* and was interviewed by Eamonn Holmes for the Channel 5 programme *How the Other Half Live*.

Andrew began his finance career at SBC Warburg in the late 1990s. Since then, he has held various senior equity roles at leading investment banks, both in London and New York. In that time, Andrew has met with the senior management teams of over one thousand companies and with hundreds of professional investors and has regularly been involved in high-profile stock market transactions. These have included the Kingdom of Sweden's sale of Nordea Bank AB in 2013 (totalling $7.6 billion) and the stock market flotation of several dozen companies including the likes of easyJet, Burberry, Campari, Carluccio's, the Carbon Trust and lastminute.com.

From January 2015 to June 2021, Andrew was a partner at an investment bank specializing in biotechnology and life sciences, WG Partners LLP.

Andrew lives in Hampshire, England, with his wife, Rachel, and their two small children, Ella and Oscar.

Introduction

'On what principle is it that with nothing but improvement behind us, we are to expect nothing but deterioration before us?'

Thomas Babbington Macaulay

This book has an unashamedly lofty ambition: I'm going to make the case that the biotechnology and related industries will have a very significant and positive impact on all of us – in the relatively near future and for many decades to come. I'm going to share a thoroughly empowering and positive way of looking at the world and one that will serve as an antidote to the epidemic of negativity so prevalent at the moment.

First, though, let me clarify what I mean by 'biotechnology' – or just 'biotech' for short – and how I intend to use those words throughout the book. *Chambers Dictionary* provides a reasonably comprehensive definition of biotechnology as:

> the use of living organisms (e.g. bacteria), or the enzymes produced by them, in the industrial manufacture of useful products, or the development of useful processes, e.g. in energy production, processing of waste, manufacture of drugs and hormones, etc.

As you can see, biotechnology is about much more than just healthcare – even though it's probably fair to say that when most people hear the word they assume it refers to the development of drug therapies. That is certainly a huge focus for the industry, but this book embraces a much broader definition and a wider and more fundamental set of themes.

Using living systems and organisms to make things that are useful to us goes way beyond curing disease alone. It is also something human beings have been doing for thousands of years, but we are on the cusp of a step-change for all sorts of exciting structural reasons. The development of the biotech and related industries will have huge consequences for human life, with everything from agricultural production and clean energy generation to the development of more powerful computers at stake. It's no exaggeration to say that biotech will be crucial to feeding the world, enabling the developed world to continue with its way of life, making that way of life possible for people in the developing world, and doing all of the above without pushing our environment to the brink.

So when I mention 'related industries', I am referring to life sciences and the medtech, healthtech, agritech, nanotech and cleantech industries, as well as things like artificial intelligence (AI), machine learning and even quantum computing. Many of these would more normally be categorized as pure 'tech' industries or sectors, but 'convergence' is a key part of the message. Increasingly, it is the *combination* of all these various technologies that gives them their potential. The 'biotech' bit, however, has a key role for two primary reasons:

- First, many of our most important remaining challenges as a species concern *biological systems*.
- Secondly, biotech is growing very fast. Biotech 'exponentials' have the potential to be even more powerful than tech exponentials.

It may not be strictly correct to use the word 'biotech' as a catch-all, but I think you'll agree it makes things rather easier. 'Our Future is Biotech/Life Sciences/Medtech/Healthtech/Agritech/Nanotech/Cleantech/AI and Quantum Computing' would have been a rather cumbersome title after all.

'Biotech', defined in this way, could be where the answers to many of our most pressing challenges as a species lie. That's why it's so important.

And crucially, biotech is going to create an enormous amount of economic value as a natural function of solving these challenges. There is a good chance that the next raft of trillion-dollar companies will emerge from these areas: the Apples, Amazons, Googles and Microsofts of the next few decades will be biotech companies. They will be the organizations that solve our most intractable problems: everything from cancer, dementia, diabetes, obesity, and numerous mental health challenges within healthcare, to (clean) power generation, agricultural productivity and environmental stewardship. Their development will mean that many more of us can live far better, safer, healthier, wealthier, happier and longer lives.

It has been estimated that up to 60 per cent of the world's physical inputs could, in principle, be made using biological means, while up to 45 per cent of the world's disease burden could be addressed using science that is conceivable today, leading to US$2–4 trillion globally of annual direct economic potential using biological applications by 2030–40. This more than likely underestimates the reality considerably. Global pharmaceutical revenues alone were already more than $1.42 trillion by the end of 2021, up from $390 billion 20 years earlier. Within that 2030–40 timeframe we could be using 'photosynthetic microorganisms' and 'biophotovoltaic' cells to make electricity. There is a reasonable chance we could be using nearly carbon-neutral 'algaculture' to make fuel for the automotive and aviation industries. We may have revolutionized the packaging industry with biologically based entirely biodegradable products to replace plastics. We may even be using 'biological' computers which are some way more powerful than the transistor-based ones in use today. It is also more than likely that biological processes will have revolutionized agriculture and the food industry.

Few people have much of an idea of the phenomenal progress being made in these areas, even now. There are already 'miracle cures' for several diseases, with far more to come soon. Not that long ago diabetes, for example, was so often a death sentence. The prognosis for millions of diabetics has been revolutionized since the development of human insulin in the early 1980s, arguably one of the first 'biotech' drugs. Giant steps have been made in the treatment of so many other diseases as well, as we shall see.

Great progress is being made in many other areas of scientific endeavour, too much of which goes sharply unreported. *Exponential* progress will also very likely drive the price of such things down far enough to make them widely available globally, not just in the 'rich' developed world.

Why this book now?

In the pages that follow I will give compelling evidence for the broad claims I've made above. My hope is that you, the reader, will derive several tangible benefits by the end of proceedings as a result.

First, you will have the ability to make better use of some of these technologies to improve your life, particularly *your health* and *mental health*. One of the sad realities when it comes to health is how poorly distributed 'best practice' and 'best information' are globally. We have never had access to more information than we do today, but a great deal of that information isn't good enough, particularly given the rise of social media and short-form content displacing long-form. There are also a number of inherent reasons why medical practitioners can take decades to implement best practice from somewhere else in the world. This happens in modern, technologically advanced nations, not just in the developing world. We will look at why this is.

Secondly, I am firmly of the belief that we should be giving consideration to this broad story when it comes to our *investments*. This is my background, after all. My first book, *How to Own the World*, is all about how important it is that *everyone* should be thinking about investment, *without exception*. I make the case that it is truly one of the great tragedies of our time that so few people think about such things enough or, indeed, at all – much to their detriment and to the detriment of society more broadly.

My strong belief is that widespread effective financial literacy, and the considered use of investment products as a result, can be a kind of 'silver bullet' – for individuals, and, more broadly, for societies as a whole. At the individual level, every person who learns enough about financial markets to become properly financially literate and optimize their financial affairs as a result, very significantly increases the chance that they will become wealthy – and almost no matter how much they earn – over time at least.

This reality has two powerful knock-on effects for society more broadly. First, every person who does this is likely to need far less state support – for them and for their dependants. This is good for public sector balance sheets which are horribly challenged all over the world. Secondly, by virtue of becoming investors, they are helping to provide capital for companies seeking to solve real human problems and/or deliver human wants and needs – just the sort of companies working in the areas which are the focus of this book.

A depressingly small number of people know anything at all about financial markets or make use of them in an effective way, even in modern liberal democracies, even with the phenomenal access to information that most people have these days, and even given the fact that the fundamental 'technology' of financial markets has now been with us for more than two centuries. This makes people's lives immeasurably harder than they might otherwise be. It is also a massive challenge to national balance sheets all over the world and to companies that need capital to do any number of fantastic things, and this includes many of those working in the biotech industry.

Having at least some investment exposure to the themes this book focuses on may well prove to be a good thing to do in the years ahead, given the enormous value that these companies are likely to create.

The third and, I would argue, most important tangible benefit that I hope you will derive from this book is nothing less than a sunnier disposition and a *material improvement in your outlook, mood and conception of the world we live in.* The ideas laid out in this book could well just put a spring in your step.

Sir Arthur C. Clarke, the British author who gave us *2001: A Space Odyssey*, famously stated as one of his Three Laws that 'Any sufficiently advanced technology is indistinguishable from magic.' There is a great deal of 'magic' going on out there, for those willing to seek it out. Science fact increasingly looks like science fiction. That magic is also highly likely to create several trillion 'currency units' of *real* economic value in the next few decades, to our great benefit as a species. It is the people at work in this sector who will be instrumental in ensuring the continuation of the unimpeachably fantastic progress our species has made in the last few centuries into the next few.

In this book, I will make the case that the world is already the best it has ever been for humanity in aggregate. Notwithstanding the daily deluge of negativity that comes from our press about pandemics, the environment, the machinations of politicians, corruption, terrorism, violence and conflict, the reality is that human experience for the vast majority of people has been steadily improving for at least two centuries – in terms of everything that actually matters: longevity, health(care), food, shelter, warmth, energy, light, literacy, leisure, travel, political, religious and sexual freedom, and extraordinary freedom from violence, war and homicide for the vast majority of the global population, as against the norm in every other previous age notwithstanding current crises in places like Ukraine, Israel and Gaza, and Yemen.

The incredible progress made by our species in the time since the Agricultural and Industrial Revolutions and the dividends paid for many billions of us alive today cannot be understated. In the main, that progress has been delivered by technological development – by the 'tech' industry if you like. Many of the remaining challenges we face will be addressed by the (bio) tech industry, if only due to the inherent nature of many of those challenges and the extent to which they involve *biological* systems.

All of this will create very significant economic value and pay real dividends for every aspect of our lives and for our impact on the planet too.

Some biotech 'magic'

One of the singular privileges of my time working in the biotech sector has been getting to meet the senior management teams of a large number of companies face to face and being able to visit their facilities and speak to their commercial staff and scientists. Over and above that, I have also been able to attend industry conferences and events and maintain a regular dialogue with professional investors, many of whom are highly specialist and incredibly knowledgeable as a result.

Time after time, I find myself stumbling across things which, to me at least, are really quite extraordinary and where science fact really does look like the science fiction of only a few years ago. Here are a few examples, in no particular order:

- An Australian company, QBiotics, that has spent more than 20 years going into the Australian rainforest looking for natural compounds which might have anticancer, wound healing and antimicrobial properties. Its lead drug can be injected directly into any tumour that can be biopsied – that is to say, the significant majority of tumours. So far, it has shown an astonishing ability to destroy those tumours and to achieve exceptional on-site wound healing thereafter.
- A highly innovative British company, Oxford Biomedica, has been a world leader in a particular kind of biotech manufacturing process for more than 20 years. In that time, it has focused on scaling up and injecting as much efficiency into that process as possible, with much larger batches, clever software, automation and robotics, for example. Doing so has meant that it has been able to take the manufacturing cost of a particular cancer treatment down tenfold, from a level that was too high to make that treatment commercially viable, to one which meant that Novartis – the company behind that treatment – was able to launch the drug. In the fullness of time, the company has said publicly that it may be able to take the cost down another five- to tenfold.
- A Belgian company, Ablynx, which has used antibodies from llamas (yes, llamas!) 'to create new therapeutics and refine existing treatments'. The company spun out of a Belgian research institute in 2001 and 20 years later was acquired by French pharmaceutical giant Sanofi for just shy of $5 billion. There are other companies elsewhere in the world doing similarly innovative things, including one in Australia which is using, of all things, shark antibodies!
- Another British company, Avacta, with a similar technology (although using human proteins rather than anything from a

llama or a shark this time) may be able to use that technology to deliver a kind of highly targeted chemotherapy. That is to say that it may be able to manufacture a chemo drug which is released only when it arrives in the tumour microenvironment, thus reducing the horrendous side effects usually associated with chemotherapy very significantly, quite possibly even getting rid of them altogether. Eventually, its technology may be able to develop a simple oral pill for cancer with no side effects, as crazy as that may sound.

- A Nobel Prize-winning technology, CRISPR, means that it is now possible to change the DNA of animals, plants and microorganisms with extremely high precision. CRISPR has had a massive impact across the life sciences, is contributing to new cancer therapies, and may eventually be able to cure inheritable diseases.

We are in the process of very significantly improving our understanding of the extraordinary complexity of microorganisms – the bacterial 'biome' and viral 'virome' – in each of us and in our environment more generally. This could be described as a new frontier of sorts and will have far-reaching consequences for health and healthcare in the years ahead.

There are companies developing wearable devices such as wrist watches and even rings which can measure key aspects of our physiology with exceptional accuracy. Such technologies can then communicate wirelessly with clever software embedded in apps in our smartphones. We are on the cusp of a step-change in 'wearable' tech and development of the idea of 'the quantified self' which will very likely significantly improve our health and, happily, go some way to shifting the focus of our healthcare systems far more towards prevention than cure – something which I believe will be an unalloyed good. Leading US entrepreneur Peter Diamandis has described this as a move from 'sick care' to 'continuous healthcare'.

One of the most exciting potential uses of biotech in agriculture is that of cell-cultured foods – the development of meat cultivated from cells rather than animals. We may be only a decade or two away from being able to bio-engineer all kinds of animal protein so accurately that it will look, smell and taste precisely the same as the beef, lamb, chicken, tuna or lobster that many of us enjoy today, but will have been grown in culture in a lab – quite possibly only a mile or two away from where we live. This approach will be infinitely more efficient, humane and positive for the environment than our current methods of securing such animal protein by growing and then slaughtering billions of animals each year and shipping meat hundreds or even thousands of miles.

At the most extreme end of what may be possible, there are even scientists who view ageing as nothing more than a disease that may be curable. British biomedical gerontologist Aubrey de Grey stated publicly as long ago as 2008 that he believes that the first human being to live to one thousand years old is probably already alive today. While his stance is extremely controversial, there are plenty of entirely sensible scientists at leading research institutions who believe that we may be on the cusp of another step-change in life expectancy, building on the three-to-four-decade improvement we have already seen in the last century or so. Happily, such scientists expect us not just to live longer but also to remain younger. We may not be too many years away from 70 or even 80 being 'the new 40' that 50 and 60 have already become (that is, if you're to believe the pages of plenty of lifestyle magazines and blog posts in recent years).

But just as there is the potential for such technologies and businesses to make an extraordinary contribution to our future, there are some very particular structural challenges faced by many of the companies working in the sector that are slowing progress.

Outside of the USA, many innovative biotech companies face an extremely difficult funding environment. There is too little capital willing and able to support companies which could be changing the world and significantly improving our lives. This is compounded still further by the fact that there are too few journalists willing to write about what they are doing.

Of course, these two things are related. When it comes to the innovation we might be making, we are fighting with one hand tied behind our back. We could be moving a great deal faster, were this not the case.

About this book

Over the course of this book, you'll discover how biotech is going to change all of our lives for the better.

Part 1 shows how the industry has already delivered some incredible science and created several trillion dollars' worth of real wealth in a relatively short period of time, and looks at the factors which have driven that outcome. It then goes on to examine some of the challenges facing the industry, particularly with respect to capital markets outside of the USA, before looking at why it is likely these challenges will be overcome.

In Part 2 we will look at the development of modern medicine, how our antibiotic and vaccine technologies have impacted infectious disease, and the relationship between this reality and the subsequent explosive growth in a worrying number of 'diseases of modernity' and disorders including diabetes, epilepsy, inflammatory bowel disease and irritable bowel syndrome, obesity, and a raft of other autoimmune diseases and mental health problems. We look at the key role played by the microbial world in all of this and at how biotech is best placed to deal with these various challenges and deliver the technologies and healthcare systems needed to revolutionize how we approach medicine at a fundamental level.

In Part 3 we look in more detail at the development of the pharmaceutical and biotech industries from penicillin to DNA, RNA and mRNA and at the rise of biologic drugs and cutting-edge innovations such as gene and cell therapy, stem cells and gene editing. We then give consideration to the key diagnostic and analytical tools which have facilitated all of the above and look at the role the industry has to play 'without us' in areas such as clean power generation, agriculture and processing power and even at the undeniably radical idea that ageing may be nothing more than a disease that could be curable, to a certain extent at least.

Much of the above may seem like 'science fiction' or even just 'wishful thinking'. I very much hope that by the end of this book, however, you'll be able to see how 'real' such things are. Biotech can and will change the world and your life in the years ahead. Happy reading – and happy investing too!

PART I

The biotech industry

Changing the world and creating wealth

In this section, we'll look at how biotech is already a multitrillion-dollar industry, with much more to come, and at why the Apples, Amazons, Googles and Microsofts of the next few decades will be biotech companies. We'll examine six powerful factors driving the industry and also some of the challenges facing biotech companies, especially outside of the USA. As we'll see, these challenges should be overcome.

The future is bright... and biotech.

I

Six structural factors driving the biotech revolution

In the time since the biotech industry first emerged, roughly from the 1970s onwards, the industry has created a great deal of economic value. By the end of 2022 the top 700 or so biopharma companies in the world were collectively valued at just shy of $5.4 trillion. There are thousands more companies not captured in that number so the real value creation is larger still.

In this first chapter of the book I thought it might be instructive to look at some of the key reasons why this has happened and, even more importantly, why it seems likely that this will continue and even accelerate in future.

Momentum is important in growth industries like biotech. More often than not, success begets success. Notwithstanding this fantastic value creation, biotech is still pretty poorly understood and poorly funded in many parts of the world. The USA and, more recently, China account for a disproportionately large percentage of the industry, which presents a significant opportunity, as other regions should eventually 'catch up' for all sorts of structural reasons.

There are six key structural factors that have been driving the biotech sector, which we'll look at in turn:

1 The science
2 Demographics
3 Equity fundamentals
4 The ethical case
5 An improving regulatory environment
6 The public sector and government.

1 The science

First, the science being done by biotech companies is getting to a point where reality increasingly looks like science fiction. Exceptional technological progress is being driven by three related and complementary exponentials:

- The fall in the cost of processing power (the impact of Moore's law)
- The fall in the cost of sequencing a human genome
- The increased ability of scientists, clinicians and regulators all over the world to collaborate using these exponentially improving technologies.

Moore's law is one of the most important exponentials in the world. This is the basic idea that the processing power of our computers per dollar spent doubles every two years or so. This phenomenon was identified in 1965 by Gordon Moore, one of the founders of leading chip manufacturer Intel Corporation. This rate of growth has persisted for more than 120 years – utterly astonishing when you think about it. Top US VC investor and Tesla ex-board member Steve Jurvetson has described the graph (Figure 1.1 below) as 'the most important graph in human history'.

As the underlying driver of essentially all of our technological progress, Moore's law has been the single most important factor for human progress overall for more than a century. Amazingly, the cadence at which it has developed goes back several decades *before* Gordon Moore came up with the idea. Even before we had transistors, microchips and computers, human ingenuity was developing at this sort of pace.

It is likely that this will continue to be the case. In my lifetime the processing power of our computers per 'currency unit' invested has multiplied as much as a billion times. Moore's law is an utterly extraordinary thing, a testament to the phenomenal ingenuity of our species and nothing less than a source of great hope for our future.

That said, there is a chart from the biotech industry that has been leaving it for dust for more than a decade: the fall in the cost of sequencing a human genome.

Figure 1.2 shows that the cost of sequencing a human genome – reading our genetic code in full – has fallen from $100 million to less than $1,000 in fewer than 20 years. The real picture is actually even more astonishing. It is difficult to estimate the precise cost of sequencing the first ever human genome, but it was thought to be around $3 billion in today's terms and took many hundreds of researchers 13 years to complete. Today there are companies that can do it in a few hours for as little as $200.

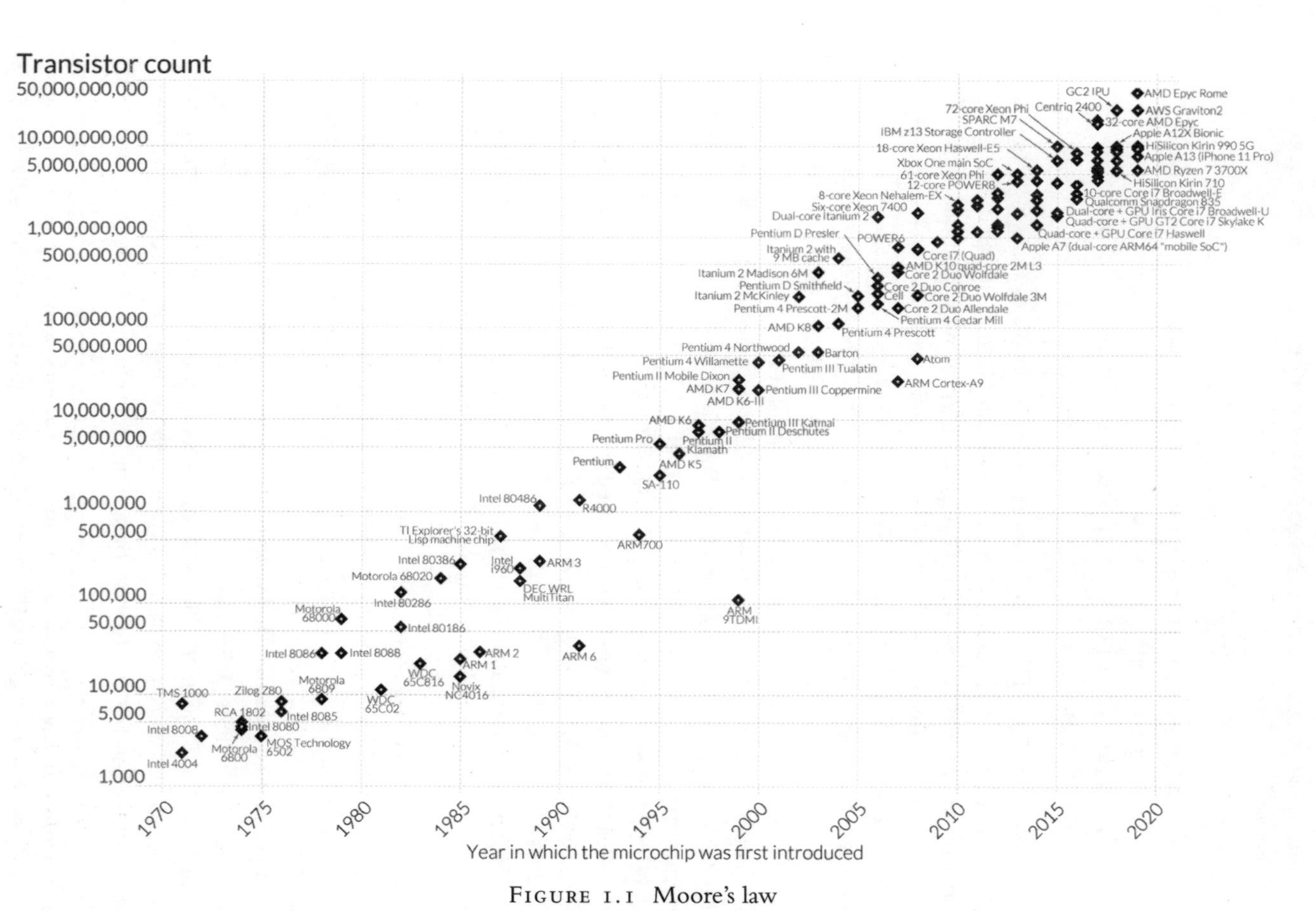

FIGURE I.I Moore's law

Source: Hannah Ritchie and Max Roser, https://ourworldindata.org/moores-law, licenced under CC.

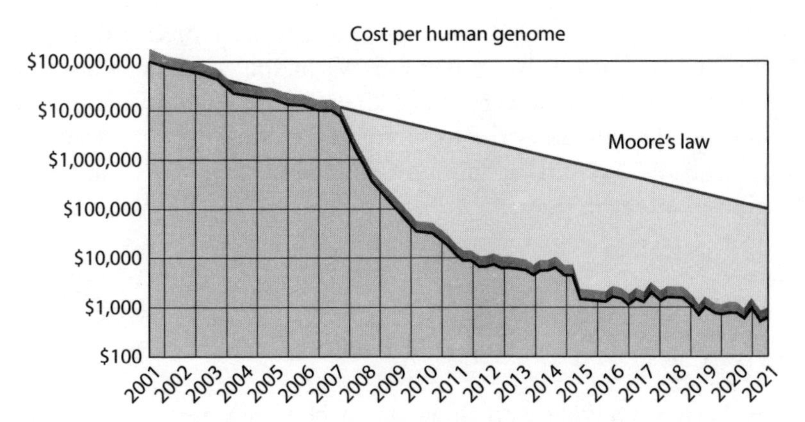

FIGURE 1.2 Cost of sequencing a human genome, 2021

Source: National Human Genome Research Institute

From 13 years and $3 billion to a few hours and $200: this is a statistic that far too few people know anything about. Importantly, this has far-reaching implications for our ability to fight disease and potentially to find effective cures for things like cancer, diabetes, Alzheimer's and dementia, and, eventually, for any and every other disease known to humanity, as hyperbolic as that may sound. It will have a role to play in many other areas of human progress, too.

As science writer Melanie Senior has put it: 'The explosion of gene editing and delivery tools, visualization methods and computer-powered data analysis is allowing scientists to tame biology in ways that weren't possible just ten or even five years ago.' Exponential rates of progress can also be seen in places like the cost and efficiency of solar power, quantum computing technology, engine efficiency, chip design, battery technology, and so on. This reality will very likely have extraordinarily positive and powerful consequences for us all – and sooner than most people realize – particularly as such technologies converge.

There is no question that scientific progress here is nothing less than phenomenal. Innovation is developing exponentially, and reality increasingly looks like science fiction. Not only is Moore's law 'the most important graph' in human history, but it is also the most important underlying investment theme, too.

At the time of writing, the main US stock market index, the S&P, has delivered average returns of roughly 10 per cent *per annum* for more than a century (going all the way back to January 1923). Moore's law has been the most fundamental driver of that fact. When you zoom out from all of the noise around financial markets and 'the news', it is the development of Moore's law that underlies the extraordinary increase in real wealth and living standards enjoyed by most of humanity since the Agricultural and Industrial Revolutions. As it delivers exponentially, this will continue and likely even accelerate into the future.

2 Demographics

While the science is developing at such a blistering pace, the need and demand for healthcare is exploding for a number of powerful structural reasons. All over the world, populations are ageing, getting dramatically more obese and suffering from a raft of other 'diseases of modernity' – and all while getting wealthier, too.

Most diseases are diseases of age. Approximately 80 per cent of cancer cases are diagnosed in people over the age of 55, and the risk increases steadily as we get older. This is also true of many of the other major diseases. As a result, as populations all over the world live longer, their demand for healthcare increases. The growth in the percentage of the world's population who are over 60 that has occurred in the last half century and the extent to which this is forecast to become even more pronounced in the future is extraordinary. You can see a visual representation in the United Nations World Population Ageing Highlights 2017 at page 4: https://www.un.org/en/development/desa/population/publications/pdf/ageing/WPA2017_Highlights.pdf.

The same is true of obesity. In most developed economies today, anywhere from half to two-thirds of the population are classified as obese or overweight. The developing world is doing its best to catch up, sadly. Just as with age, obesity significantly increases rates of illness. It is estimated that there are now more than half a billion people in the world with diabetes, for example – a fact that has much to do with those rates of obesity.

Both of these factors have driven a massive increase in the demand for all kinds of healthcare products and services, and these factors are getting stronger every year. The world's largest specialist diabetes-focused business, Novo Nordisk, is now valued at well over $350 billion, for example, and it has been estimated that oncology spending was more than $200 billion in 2023 alone.

Humanity is also increasingly suffering from a large number of 'diseases of modernity' and disorders, including, but not limited to, debilitating allergies such as hay fever, nut allergies, eczema and asthma, epilepsy, inflammatory bowel disease and irritable bowel syndrome, and a raft of other autoimmune diseases including rheumatoid arthritis, coeliac disease, myositis and lupus. In addition, rates of depression and other mental illnesses are also on the rise all over the world. We will look in more detail at why this has been happening later in the book. For now, however, it is enough to highlight that these are all significant market opportunities for businesses working in the sector.

It is also the case that wealthier populations spend more money on healthcare than poorer ones – because they can afford to. It wasn't that long ago that most of the world's population had no access to healthcare whatsoever. Decent healthcare was the preserve of a privileged few in the developed world. This has been changing rapidly. Today, billions of people all over Asia and Latin America and even some of the wealthier parts of Africa have access to functional healthcare services and drugs in a way that their grandparents or even their parents did not. This phenomenon is also developing exponentially.

The Chinese healthcare market is already the second biggest in the world, for example, having grown at an eye-watering pace for many years. A number of leading market commentators believe that it will overtake the US market in the reasonably near future. This fact alone is likely to be bullish for companies with quality assets which can address that market, let alone so many other large markets such as India, Pakistan, Indonesia and most of Africa in the fullness of time.

3 Equity fundamentals

This is a more 'specialist' point than the two I've made so far and a bit less intuitive for someone without training in financial analysis, but successful biotech companies exhibit a number of attractive 'equity fundamentals'. These include: significant pricing power, high margins, long product cycles and high barriers to entry thanks to their long-duration intellectual property (patent protection).

Perhaps more important from a pure investment perspective, they're also sitting in a sector with unprecedented financial firepower. In 2023 leading accountancy firm Ernst & Young (EY) estimated that there was around $1.4 trillion of cash available to biopharma and medtech companies for investment and mergers and acquisitions (M&A). There is a similarly large sum in the private equity and venture capital sectors focused on funding pharmaceutical and life science companies.

What this means is that there is an astonishing quantity of money available across the world to develop the science and, potentially, to bid up the prices of the companies involved. The sheer scale of the investment capital focused on and available to the sector implies that there are likely to be plenty of potential buyers for any company that can demonstrate it has something valuable.

This reality is stronger in healthcare and biotech than in nearly any other economic sector – given the sheer size of these financial firepower numbers and given the fact that an increasing number of the largest companies in the sector seek to acquire or license innovative science and intellectual property rather than develop it themselves.

4 The ethical case

A fourth key structural argument for the merits of the sector concerns the growing focus on so-called 'ethical' investing across much of the world.

You may be aware that one of the biggest investment themes for the last several years has been something called ESG. ESG stands for 'environmental, social and governance'. In 2019, 2020 and 2021 ESG funds attracted $285, $542 and $649 billion of investment capital, respectively. This was not far off half of all new investment flows for the last two of those years in Europe.

The basic idea is that the more 'ESG' a company or fund is, the more 'ethical' you can consider it to be across those three broad areas:

- **Environmental** – This is perhaps the most 'obvious' of the three components. An ESG assessment will look at things like a company's impact on climate change, energy consumption, resource depletion, pollution, disposal of hazardous waste, deforestation, impact on biodiversity, animal welfare and so forth.
- **Social** – Here the ESG assessment looks at a company's impact on all of its stakeholders as well as on society as a whole – put simply, a given company's impact on people. This would

include things like conditions for employees: fairness, diversity, labour rights, working conditions and so on. In the biotech and healthcare setting specifically, patient outcomes would obviously be a key consideration.

A company scoring well here would also demonstrate that it was mindful of the same factors at work with its suppliers as well as for its own staff. It might be great to work for a Californian tech company in Los Angeles or a British fashion brand in London, for example, but not so great to work for the Chinese company supplying that tech company with components or the Sri Lankan company supplying the fashion brand with clothes. A company should only score well here if it can demonstrate it is thinking about all of the above across its entire supply chain.

- **Governance** – This aspect of an ESG assessment examines how well a company addresses things like the use of accurate and transparent accounting policies, executive pay, and its treatment of shareholders, particularly in terms of their voting rights and their ability to exert control on a company's management when needed. Governance is also concerned with potential conflicts of interest, the choice of board members, or where a company might make political donations. Interestingly, big tech companies, which might otherwise seem like quite enlightened 'good actors', often perform pretty badly here given concerns over things like data misuse, tax avoidance and higher voting rights for founders' shares versus everyone else's.

ESG will also want to take account of a company's involvement in 'sin' products such as alcohol, tobacco and guns (or other armaments and defence products more generally).

A key point to make about ESG as a whole is that it is a complicated and subjective area. It is hard to be black and white. The ESG credentials of any given company or investment are inherently difficult to establish. Information about a big company will often be highly imperfect and, even if you can get accurate information, there is a huge element of subjectivity and judgement when 'scoring' such things. By extension, it is even harder to establish the aggregate ESG credentials of a fund or investment company which may own hundreds or even thousands of companies.

Take the environmental bit, for example. Copper requires a great deal of energy to mine but there could be no electric cars or wind turbines without it. An oil company is very clearly a 'carbon' company, but how might we legislate for, say, a large medical logistics company delivering life-saving products but using a fair bit of diesel and aviation fuel in doing so? The more you think about how interconnected everything is, the more bewildering any judgement becomes and the harder it is to draw hard-and-fast conclusions about a company's environmental credentials.

Much the same could be said of the 'social' and 'governance' components. During our careers many of us will have had direct experience of working for a company with a fantastic reputation and brand while marvelling at just how rotten and imperfect it is when you're on the inside. Even if you work for a wonderful company that you love and that you're very proud of, how many times might you have heard your friends and peers tell stories about how hopeless their company is and say things like 'If people only knew the truth and what a disaster we are'? I've lost track of how many times I've come across such sentiments in more than 20 years of working with people at some of the 'greatest' companies in the world.

That said, many, possibly even most, companies in the world, certainly stock-market-listed ones, are becoming naturally 'more ESG' over time for several inherent, structural reasons. Basically, because most of their stakeholders are *forcing them to*.

There is often an inherent micro-push at many companies in the world towards being better corporate citizens, because of the impact of the people who run those companies, work for them, buy from them and invest in them. Not that long ago, many companies were run by 'rich, old, greedy, white men'. Many of the investors in those companies were cut from the same cloth, and relatively few of their employees or customers thought that much about whether a company was exploiting minimum-wage workers or damaging the natural world.

This has changed very significantly in the last 20 years or so. There is therefore a 'supply' push element to this, in that a very significant percentage of companies are actually run by people who care about all things ESG just as much as the rest of the population might. There are plenty of companies where management and employees have a real sense of mission about such things. I have seen this first-hand in the healthcare industry time and time again.

So often I see people ranting about the behaviour of big pharma companies on social media. While there are always bad actors in any area of human endeavour, I have found that CEOs, divisional management and research staff in most companies with whom I've worked directly are invariably driven far more by a passionate desire to improve patient care, or even the world at large, than they are by greed or self-interest. I think it is also important to note that what may have been the status quo in the past is not necessarily true today. It is pretty easy to find examples of 'bad pharma'. Entire books have been written on the subject after all. But things are improving and will continue to improve. Bad news gets more attention than good news – a subject to which we will return.

Even where management teams might not be inclined to be good corporate actors out of personal choice, there is also a powerful 'demand pull' factor at play here, in that they are effectively forced to be good actors or face real problems by the very fact that they need to hire young staff, attract capital and sell their products to consumers who care about such things more than ever before.

You can see this in popular culture: Julia Roberts won the Best Actress Oscar in 2001 for her portrayal of US activist Erin Brockovich. The film of the same name told the story of how Brockovich took on the Pacific Gas and Electricity Corporation, one of the biggest utility companies in the world, and secured $333 million of compensation for nearly 200 people who, the case found, had been poisoned by the company over many years. This was a real wake-up call for corporate America and illustrative of just how much the mood music had changed by 1996 when that court case took place. This sort of outcome had rarely happened previously. Erin Brockovich could never have secured that result in the 1950s, 1960s, 1970s or even 1980s, and since then this reality has been supercharged by social media, of course, even if somewhat imperfectly.

Above, I have suggested there is a micro-push towards ESG at the level of the company. There is a fair bit of evidence that there is also a related top-down macro-pull. That is to say that, increasingly, the world's biggest and leading companies are unilaterally deciding to put ESG centre-stage in much of what they do.

There are myriad examples of this. Unilever is one of the biggest consumer products companies in the world. In 2022 it sold just over €60 billion worth of many of the products we use every week in 190 countries. Its brands include Dove soap, VO5 hair products, Hellmann's mayonnaise, Ben & Jerry's, Cornetto and Magnum ice cream, Lipton tea, Domestos bleach and Persil washing products – the list is very long.

In June 2020, the company announced 'a new range of measures and commitments designed to improve the health of the planet by taking even more decisive action to fight climate change, and protect and regenerate nature, to preserve resources for future generations'. It underpinned this with an ambitious target: 'Unilever will achieve Net Zero emissions from all our products by 2039.' The company also set up a €1 billion 'Climate & Nature Fund'.

In the investment industry BlackRock is one of the biggest investment companies in the world, quite often *the* biggest on any given day. At the time of writing, it is in charge of more than $9 trillion of client assets. In 2020 its CEO, Larry Fink, wrote a letter to thousands of CEOs around the world entitled 'A Fundamental Reshaping of Finance', announcing 'a number of initiatives to place sustainability at the center of our investment approach'. Even the nasty old oil and gas industry seems to be subject to the ESG macro-pull: again in 2020 Shell CEO Ben van Beurden announced the company's intention to be 'a net-zero emissions energy business by 2050 or sooner'. And Shell is by no means the only massive company in that sector with similar aspirations. It is often forgotten, for example, that the oil majors are some of the biggest investors in clean energy technologies, committing $9 billion in 2020 alone. This is not surprising when you see headlines such as Yale Climate Connections' 'Institutional Investors are running away from big oil ...' – precisely the point I am making.

There are thousands more examples from companies all over the world. There is much talk of 'greenwashing' when these sorts of companies make these sorts of announcements, and there is certainly a fair bit of that to be sure, but the mere fact that they even feel that this is something they need to be doing is a real improvement on times past.

The biotech and healthcare industry generally scores very well on all things ESG. Leading Danish reputational intelligence company RepTrak ranked Pharmaceuticals, Biotechnology and Life Sciences as the number-one industry sector for ESG in 2021. Animal testing remains a challenge for companies working in the industry, and this does negatively impact their ESG scoring, but this is more than compensated for by the broader impact on patient outcomes. It is also perhaps worth mentioning in passing that there are a reasonable number of sensible scientists who believe the days of animal testing may be numbered. There are alternative technologies showing great promise in that respect.

Whatever may or may not happen with animal testing in future, the biotech industry can be a big beneficiary of the growth in ESG investing, and this theme will more than likely be durable and continue long into the future. In 2019 US investment company Fidelity published a study showing that 77 per cent of wealthy millennials have made an 'impact' (ESG) investment. Another investment firm, American Century, has estimated that one-third of such investors would make healthcare their number-one priority were they to make an impact investment. There is a great deal of capital seeking to support innovation here.

I'm not saying that the world is perfect and that every company now behaves extremely well – but the direction of travel is encouraging. The behaviour of corporate entities will necessarily mirror the society we live in, because everyone involved with those companies comes from within that same society.

The mere fact that ESG even exists and nearly $1.5 trillion of new investment money flowed into ESG funds in recent years is a pretty strong indicator that this is the case.

5 An improving regulatory environment

Estimates vary, but on average, getting a drug to market costs nearly $3 billion and takes about ten years. This enormous cost and time requirement has a great deal to do with the regulatory requirements of getting a drug approved – by organizations such as the Food and Drug Administration (FDA) in the USA, the European Medicines Agency (EMA) in Europe, and the National Medical Products Administration (NMPA) in China. This is one of the biggest challenges facing biotech companies working in healthcare – particularly small ones that are less well equipped to be able to deal with the enormous costs and time required before it is possible to actually make any money.

Anything that happens to speed up this process and/or reduce costs is highly beneficial for the value of the sector. Happily, the direction of travel in this respect in recent years has been positive. Regulators all over the world are better funded, politically empowered and increasingly comfortable with complicated 'bleeding-edge' science.

In the early days of the emergence of the biotech sector, the regulatory agencies didn't have the staff needed to understand what the leading biotech companies were doing. As new ground was broken scientifically, particularly from the 1980s onwards, the cutting edge of the industry was often really quite far ahead of regulators in terms of the technology. The scientists at such companies were the sorts of people winning Nobel Prizes, after all. Understandably, this made those regulators pretty reticent about approving their products. It is hard to be comfortable approving a drug product if you don't understand it.

This reality took several years to resolve itself and likely increased the cost and complexity of seeking regulatory approval for much of the market in that time. When we look at the overall costs and time required to approve a drug, over the last few decades at least, we should give consideration to this fact and hope that the experience of the past may not be that relevant to the future – and that average timelines and costs may come down.

This reality probably isn't that surprising. Early businesses in the market for any new technology confront a significant first-mover *disadvantage* – of very significant costs, uncertainty around whether there will be market acceptance and adoption of their product (including by regulatory bodies), lack of sufficient network effects to get to critical mass, and so on – long before they can then benefit from the first-mover *advantages* which they hope will arrive eventually and deliver 'super-normal' economic rewards.

We can see this in the development of every major industry in history. Consider the development of railways in the 1800s, or of the automotive and aviation industries a few decades later. More recently, the evolution of the internet and the market for electric cars and solar power give us plenty of examples of the same thing, even when you consider the trajectory of the handful of companies that survived and 'won', such as Apple, Amazon and Google/Alphabet.

In recent years, however, the regulatory situation for healthcare has been steadily improving all over the world. Notwithstanding how fast the industry has grown, and the vast range of new frontiers being explored, many market participants seem to feel that the main regulators have the skills and the resources required to do a decent job today. They are still a bottleneck, without question, but this is almost always going to be the case with any regulatory body given the constraints on public sector funding as against the hundreds of billions that the private sector can bring to bear. The finance industry is no different from the healthcare industry in this respect, for example, when it comes to financial regulation.

As another illustration of the direction of travel here, in late 2015 I attended a presentation in London given by senior members of the Chinese regulatory authorities where one of the key themes was the investment they were making in scientific talent. In the last decade it is estimated that senior review scientists at relevant Chinese agencies have increased from fewer than 100 individuals to more than 800.

I don't think it is overly challenging to argue that this general trend has been supercharged since the first quarter of 2020 by the COVID-19 pandemic. All over the world the political imperative to find an effective coronavirus vaccine as quickly as possible focused governments and their electorates on the importance of healthcare regulation like never before. As leading British law firm Linklaters put it:

> In order to support the industry's response to Covid-19, governments have sought to release some regulatory burdens on healthcare companies and accelerate market access for important products [...] in the medium term, it is clear that there may be more change to come. Governments and regulators are looking closely at healthcare regulation, seeking to address weaknesses revealed by the current pandemic.

This acceleration of market access was a key driver in enabling the development of a vaccine in record time, alongside a raft of other structural factors to do with the development of the science and the role played by several technologies developing in tandem at an exponential rate. The need for a COVID vaccine and the metaphorical kick up the proverbial this gave governments all over the world was merely the final catalyst of a number of underlying themes which had been developing for decades. As Azeem Azhar – a British writer, podcaster and creator of the 'Exponential View' newsletter – put it in his 2021 book *Exponential: Order and Chaos in an Age of Accelerating Technology*: 'The speed with which scientists went from identifying a novel pathogen to getting a working vaccine was unprecedented in human history.'

Without wishing to open the Pandora's box of conspiracy theories associated with the pandemic, it seems safe to say that we were able to deliver vaccines in record time, not because a shadowy global elite of lizard people led by Bill Gates and the late Her Majesty the Queen were trying to control us all but, rather more prosaically, because the incredibly powerful combination of big data, computing power and things like the cost of gene sequencing enabled scientists all over the world to collaborate and progress like never before and because governments and regulators finally had a powerful political motive to move as fast as science allowed.

I should say for the sake of balance that some of the more sensible misgivings about the vaccine programmes may prove to have some merit in time. Drug development is extremely hard after all, and we are in uncharted waters here to a great extent given the unprecedented size of the patient population for the COVID vaccinations. Ultimately if you treat a billion people, the number of 'Adverse Events' (AEs) and even fatalities that will result are likely to be one thousand times higher in absolute terms than if you treat a million people, by definition.

In a world where more than 9.6 billion doses of a treatment have been given, there are going to be AEs and fatalities. Disaggregating correlation as against causation for those events is a very complicated business requiring a huge amount of data and time.

Arguably the most effective prism through which to consider such things is the idea of Benthamite utility – from the writings of the Philosopher Jeremy Bentham: that is to say that our policy response to this, and indeed to pretty much every other thing that matters, should be considered with respect to what produces the greatest good for the greatest percentage of the population, overall.

It seems reasonably likely that over time, the Benthamite case for the world's vaccine programmes will be clear – in that the steps taken in response to the pandemic, in terms of rolling out the vaccine programmes at least, were the right ones to take for the benefit of humanity in aggregate.

Whatever your views on vaccines, lockdowns and every other controversial element of the pandemic, the broad point I'm trying to make here concerns only the question of whether or not the crisis may have precipitated a step-change improvement in how healthcare regulation works.

It will be fascinating to see the extent to which these new regulatory habits could be durable. People in key industry roles seem to think that they might be. It is interesting to note, for example, that the Ebola vaccine ERVEBO was only approved in late 2019 notwithstanding the fact that work began on that technology in the 1990s. That is to say that it took the global healthcare industry and the regulators more than 20 years to get this vaccine over the line. Prior to that, meningitis took about 90 years and polio around 45. Those sorts of timelines will be less likely in future. The science and the regulators can both move far more quickly now. Dare we hope that looking back from the vantage point of a few years from now, this may end up being a very welcome silver lining on an otherwise very black cloud?

6 The public sector and government

Over and above increased funding for regulators, more generally the biotech sector has been identified as a key strategic sector by governments in many parts of the world for many years. The sector creates significant economic value, and average wages in the industry are materially higher than in many others.

The US government has long supported the sector, given its strategic importance and economic value creation. President Joe Biden's budget for the 2023 financial year included a sum of US$62.5 billion to be allocated to the National Institutes for Health, the primary federal agency in the USA responsible for conducting and supporting medical research. There are also numerous federal, state and city programmes that support the sector, over and above high-profile private sector organizations such as the Bill & Melinda Gates Foundation, the Chan Zuckerberg Initiative, the Salk Institute, the Broad Institute and Schmidt Futures and many others that are a source of several billion dollars more in funding for biotech research and development.

The EU is similarly supportive. Healthcare expenditure is nearly 10 per cent of EU GDP, after all. By 2020 the European Investment Bank had provided total financing of close to €35 billion for healthcare-related projects worldwide.

India has the largest public health insurance scheme in the world, providing 500 million people with free healthcare. In its 2021 budget the Indian government increased spending on healthcare by 137 per cent over the previous year, and described healthcare as the top priority in its national budget.

In China, government health expenditure has more than tripled since health reforms began in 2009. In October 2016 President Xi Jinping announced the 'Healthy China Blueprint', a declaration that made public health a precondition for all future economic and social development. It aims to expand the size of the health service industry to ¥16 trillion (US$2.35 trillion) by 2030.

All over the world it is relatively easy to evidence an increased focus on and investment in healthcare and biotech from governments.

In summary

Taken together, the six key structural drivers outlined above have driven very significant value creation over the last 30 years and more. In the time since its creation the biotech industry has created several trillion dollars of real wealth and provided gainful employment to well over a million scientists globally. Even more importantly, it has driven exceptional scientific progress as a result. In the healthcare setting this has impacted and improved the lives of millions of individuals living with a wide range of diseases including ones as problematic as cancer and diabetes. Beyond healthcare it is already showing that it has a role to play in a great deal else besides, whether in revolutionizing agriculture, clean power generation, or transcending physical limits to processing power.

Crucially, the value creation and scientific progress of the last few decades will continue into the next few. In fact, it will very likely accelerate. Throughout history *real* wealth creation has invariably come from solving *real* human problems. Farming and agriculture developed because we needed feeding and food security. Architecture and construction developed because we needed shelter. Our desire to travel and transport goods led to the development of the shipping, rail, automotive and aviation industries.

Many of the most intractable remaining problems we confront as a species today are all about *biological systems*. This means that it will be the biotech industry that is naturally best placed to solve those problems and be one of the foremost drivers of real economic growth and wealth creation in future as a result.

The other reason that value creation and scientific progress will accelerate is, quite simply, because there are so many key areas of innovation which are growing at an exponential rate. Later in the book we will look at areas such as gene and cell therapy, gene editing, microbiome therapeutics, and machine learning and artificial intelligence (AI). All of these exciting emerging fields, and, indeed, many others, are currently growing at somewhere between 20 and more than 30 per cent per annum and are forecast to continue to do so for some years to come. That is how fast the science and its accompanying value creation are moving.

We will come on to look at some of this amazing science. Before we do, however, it is worth first looking at some of the challenges confronted by the industry, particularly outside of the USA.

2

The challenges facing biotech companies

It is clear that the biotech industry has already come a long way, and there is even more to be excited about in future. But it isn't all plain sailing. There are some very particular structural challenges faced by many of the companies working in the sector that are slowing progress, even in some of the most technologically advanced nations in the world. Outside of the USA in particular, many innovative biotech companies face an extremely difficult funding environment. There is too little capital available to support and grow companies that could be changing the world. And there are too few mainstream journalists writing about what the industry is doing as well. When it comes to the innovation we could be making, we are fighting with one hand tied behind our backs.

Before we move on to look at a number of the most exciting elements of why 'our future is biotech', I thought it would be instructive to look at some of these challenges, if only to shine a light on them. They need to be addressed for the industry to deliver all that it might.

A regional science and wealth creation gap

The UK and Europe are home to 43 of the top research universities for biotech in the world, as against 34 in the USA. Together, British and European academics publish more than twice as much relevant research as their American peers. British and European academics have also won an astonishing number of Nobel Prizes, particularly on a per capita basis and in the most relevant category to our purposes – 'Physiology or Medicine'. In the UK, the University of Cambridge alone has won 27 Nobel Prizes for Medicine and another 25 for Chemistry, many of which have been utterly instrumental in the development of the biotech industry.

Despite this excellent science and academic prowess, the UK and Europe lag a long way behind the USA in terms of company formation and value creation. The same is true elsewhere in the world where there is similarly good-quality science, particularly in Australia for example. The UK, Europe, Australia and other places with great science can be reasonably good at supporting early-stage companies but are demonstrably bad at scaling those companies and delivering the tangible commercial successes that get that science to market for the benefit of patients and to drive real wealth creation.

There are a large number of structural reasons for this. First, outside of the USA, historically there has too often been little to no entrepreneurial culture among scientists working in academia, even in modern developed countries and world-leading academic institutions. A large number of American universities can point to a multi-decade track-record of seeking to commercialize work done by their academics, and a well-developed technology transfer and venture capital infrastructure has grown up to support that reality. In the USA, it is reasonably commonplace to find academics working in biotech- or healthcare-related fields who not only teach and see patients but who have also founded a biotech or medtech company or two or even more. This is far less common outside the USA.

In the UK, Europe and elsewhere there is far less of an entrepreneurial culture and can also be less freedom to operate. All too often the notion of looking to commercialize their work by founding and building a company doesn't even occur to an academic scientist or clinician and, even if it did, they would have no idea of how to go about it. Many academics and clinicians even view such things with a certain disdain and feel that any attempt to monetize their work is somehow improper.

There is a certain kind of purist academic scientist entirely disinclined to get their 'hands dirty' with 'grubby commerce' and who view such things as prejudicial to their ability to do 'good science'. That same stance means that there are plenty of places outside the USA where scientists working in academia can even be *prohibited* from seeking to do anything commercial by the institutions for whom they work. Although things are changing gradually, this means that outside the USA the technology transfer and funding infrastructure is far less developed.

This is further compounded by the tax system in many European countries and, related to that, how the investment industry functions, or, more pertinently, so often doesn't function, when it comes to biotech progress and value creation.

The importance of investment

No matter how good a given scientific development may or may not be, tens or even hundreds of millions of investment capital will be needed to commercialize that innovation and bring it to patients and/or customers. As we saw in Chapter 1, it can take as long as ten years and as much as $3 billion to get a pharmaceutical drug to market.

A therapeutic biotech company working on a new drug will need to navigate four distinct stages (one preclinical stage and then three clinical 'phases') in the development of that drug. First, it would need to do a great deal of 'preclinical' work on a given chemical or biological compound before it can be given to a human being. There are then three distinct 'phases' of clinical trials where the drug is given to patients. Broadly, Phase I is about checking that the drug is safe and establishing the right dose. Phase II is about checking it actually does something 'statistically significant' in a relatively small number of patients – that is, it actually works well enough versus any other treatments that are already out there in the market. Phase III is about verifying that statistical significance in a much larger number of people to prove that the Phase II wasn't just a statistical fluke. Only once a drug asset has navigated all four of those developmental stages can it hope to be approved by regulatory authorities and start generating revenues. Each of these stages requires significant investment, and a drug can fail at any one of them.

Medical devices and certain other products of the 'biotech' industry can require rather less capital than drug development, but the general point stands that (very) significant investment capital will be required to bring great science to a place where it can deliver something tangible for the world and for patients or for the clean energy or agriculture industries.

To find those many millions of investment capital, a scientist or entrepreneur will need to found a company and then grow that company through a series of funding rounds. Initially, a company will raise 'seed' capital. This will usually come from a small number of high-net-worth individuals, or 'angel investors', and perhaps a handful of professional investors, such as venture capital (VC) firms that specialize in such things.

At this point, a given biotech company would look to raise a few million to hire some people, secure premises and start the work required to take their idea to market. For a company looking to create a new drug or medical device, this would include conducting research and funding early-stage clinical trials to show that their product could work, and securing patent protection and engaging with the relevant regulatory bodies – all of which is expensive and time consuming.

Assuming this all goes well, within a few years the company will then need a great deal more money to fund later-stage clinical trials with many more patients, which is extremely expensive. In the fullness of time, they may also need yet more capital to fund manufacturing and sales and marketing.

Although there is no hard-and-fast rule about such things, ordinarily during the first stage of this trajectory the company will be a private company – funded by angel investors and venture capital. At some point, however, in order to raise the tens or even hundreds of millions required to get to some kind of finishing line commercially, a company will very likely need to 'go public' and float on a stock market by conducting an IPO, or initial public offering.

And it is here that the world outside of the USA is most particularly challenged when it comes to the funding of biotech companies. There are a number of specific structural and historical reasons for a clear lack of interest in and support for many innovative companies at this stage in their development, and these have far more to do with the way in which capital markets and the investment industry function than the potential inherent in many of those businesses, their possible commercial trajectory, and what they might be able to deliver. All too often this means that a great deal of potentially fantastic science can be strangled at birth, which is highly problematic for all of us.

We can boil these structural and historical reasons down to the following:

- Investment markets are fragmented.
- Smaller company focused investors tend to be generalists.
- Many biotech companies are too small for specialist investors.
- The market is therefore 'inefficient'.

We'll look at each of these in turn.

Investment markets are fragmented

First, there is the entirely prosaic fact that European investment markets are inherently vastly more fragmented than the US market. There are biotech companies listed on 15 different European stock exchanges (including in the UK) as compared to just the one stock market in the USA. This matters because the stock market in each European country, or in Australia for example, is therefore far smaller than the US market. There is far less 'capital depth'. This means there is far less money, in aggregate, willing and able to support companies generally and biotech companies in particular.

We like to think that investment capital is wonderfully global and mobile and that markets are 'efficient'. This is broadly true when it comes to very large companies. Most stock-market-listed businesses in the world that are worth at least several billion dollars will generally be able to attract investment from shareholders based all over the world. For tens of thousands of smaller companies, however – that is, for the majority of companies in the world – stock markets are actually still incredibly parochial and local.

Broadly, any company that is worth less than a few billion dollars (or pounds or euros) is described by the investment industry as a 'smaller company' by definition. In the main, smaller companies will need to rely to a great extent on their local investment market to secure funding. A British company valued at tens or even hundreds of millions will generally need to spend most of their time looking for support from British investors. The same will be true for companies everywhere, whether in France, Germany, Scandinavia, Australia, Japan or South Korea.

Smaller company focused investors tend to be generalists

Earlier-stage, innovative biotech companies are almost always 'smaller' companies by definition. This means that when they need to raise those many millions of investment capital to fund clinical trials, product development or manufacturing, in the main they are going to have to try to find the money on their doorstep.

And this is where the trouble begins. A key problem is that professional *smaller company* investors are invariably *generalist* investors. The people running the large pots of money tasked with giving consideration to smaller companies spend their time looking at businesses from every sector of the economy. One day they might be looking at an oil company, the next they might be meeting with a brewer, retailer or software company. Such individuals can meet several hundred companies a year.

Working out whether a company is a buy or a sell is a highly complicated, subjective and capricious game, and almost certainly more of an art than a science, but any investment professional worth their salt will be using a combination of 'valuation techniques' in a bid to help them make those decisions. Many readers will be familiar with such techniques. They would include things like P/E (price/earnings) and EV/EBITDA (enterprise value/earnings before interest, taxes, depreciation and amortization) ratios, metrics like FCF (free cash flow), ROCE (return on capital employed) and ROIC (return on invested capital), and 'gearing ratios' which help an investor understand how much debt a company uses and whether this is at an appropriate level in terms of risk and cost of capital.

The great majority of businesses can be valued using these conventional valuation metrics. A big problem for the biotech industry is that this sort of analysis is much harder, often even impossible to do. Biotech companies are uniquely *specialist* investments compared to companies from most of those other industry sectors, and such traditional valuation metrics generally can't be applied given these businesses can often be several years away from profitability or even from making any revenues.

The way investors might value a biotech company or a given 'asset' (e.g. a potential drug) is, very broadly, as a function of the perceived percentage chance that the asset has of making a given amount of money, should it make it all the way through to market. Painting with a broad brush, if a company has an asset that has, say, a 20 per cent chance of having a drug that is worth $1 billion, you might value that asset at roughly $200 million. Obviously, it is rather more complicated than that, but this is the general idea.

This is how it is possible for a biotech company to be worth millions or even billions in some cases many years before it makes any revenue or profit. If a company has a 'Phase I' drug that might do something fantastic in breast cancer, perhaps, it will be several years away from delivering on that promise commercially. But if investors looking at that company believe that they have a breast cancer drug that has a 10 per cent chance of being worth $5 billion a few years from now, then they might 'discount' that value back to that value in today's money, multiply that number by 10 per cent and ascribe, say, $350 million of value to what that company has today, notwithstanding the fact that they may be years away from making any money at all.

This entire approach is incredibly difficult and specialist. As a result, most *generalist* smaller company fund managers feel, perhaps entirely justifiably, that they don't possess the knowledge to consider an investment in such companies with complicated science that they view as being outside their natural comfort zone and which they struggle to value. At the very least, such investors have generally felt that there was much easier hunting for them elsewhere in the stock market.

This is perhaps even easier to understand when you consider the track record of many of those generalist fund managers who focus on smaller companies. Many such individuals have exceptional long-run performance track records. It is not hard to find examples of professionals in this market niche who have achieved average annualized returns in the low to mid-teens for a decade or more. This is often true of many of the people running the largest pots of money – which is why they're the ones running them!

These fund managers have been extremely successful over many years sticking to those areas of the stock market that they know well. There is little to no incentive for individuals in that position to decide to go 'off-piste' and choose to invest in companies, such as biotech companies, which they view as risky and much harder to understand and evaluate than most of the many hundred other companies that they could select for their portfolios. This is a clear and entirely understandable case of 'if it ain't broke, don't fix it' and the value for most people professionally in sticking to what you know best and what works for you personally. You might argue that this is a position that is hard to criticize.

Another challenge here is that biotech is, necessarily, a sector where it is extremely important to hold a reasonably large number of companies if you *are* going to get involved. You need to own a portfolio of companies, not just one or two holdings. Because many biotech companies lose money, if they fail to deliver their commercial milestones, by failing a clinical trial, for example, they can see their share prices go to zero or close to zero pretty quickly. For this reason, it is important to hold at least ten, preferably 20 or more such companies to mitigate that risk through diversification. Over time, this approach has a good chance of delivering a great result by virtue of the winners more than compensating for the losers, but it implies that fund managers really need to commit to the space. They need to do it properly or not at all, with all that this implies for workload and capital commitment. This is yet another reason that generalist smaller company investors tend to avoid the sector altogether. It makes their professional life a good deal easier, even if it has disastrous consequences for so many companies and for our ability as a society to develop good science optimally.

In my time working in the biotech sector in the UK, I worked out more or less empirically that, of 50 of the very largest investment companies in the UK, only *six* of them had any exposure to small UK life sciences companies, and even those few brave souls invariably had very little and very sporadic exposure. There was basically no large fund management group in the UK that had any kind of concerted focus on smaller British biotech companies or any internal mandate or desire to invest in them. Even worse, the few *specialist* biotech funds that are based in the UK tend to be almost entirely focused on larger companies and primarily companies based in the USA. The same phenomenon exists in many other markets outside of the USA.

Many biotech companies are too small for specialist investors

So, biotech companies that need to raise many millions to fund what they are doing invariably find that they are seen as too specialist for the smaller company investors in their home market who should be their natural audience and best placed to support them with investment capital.

If small biotech companies are too 'specialist' for most smaller company investors, then surely the solution to that problem is for such companies to approach the *specialist* investors who *do* understand them?

And this is where we find another example of what economists would describe as a 'market failure'. The problem is that a great deal of the money that specializes in all things biotech sits with specialist investors who are based in the USA. If you are a small British, French, German or Australian biotech company looking for $100 million to fund a potential cure for a given cancer, you might think that all you and your advisers need to do is approach those specialist investors and enthuse them sufficiently about your exciting prospects to secure that money. There are two fundamentally mundane reasons why, sadly, it simply doesn't work like that, or hasn't historically at least.

1 The 'human resource problem'

The first is what I would describe as the 'human resource problem'. Specialist biotech investors in the USA simply don't have enough time to be looking at small biotech companies from outside their own country. If you are a biotech analyst or fund manager sitting in New York, Boston or San Francisco, you will have to follow several hundred biotech and related companies that are already listed on the US stock market as a key part of your role.

On top of that, you might own upwards of 50 companies in your portfolio already, possibly some way more than that depending on the kind of fund you run. Most of those companies will report quarterly (four times a year) and even more frequently on an ad hoc basis when they have something commercially important to say. You better have a pretty decent handle on the progress every single one of them is making across a wide range of different products and therapeutic areas at any given time. If your boss turns up at your desk one day and asks you about the most recent clinical trial read out for given company 'x', you need to know the answer. This is challenging enough!

Beyond that, if you are sufficiently diligent in your role, you will also need to keep a weather eye on a large number of other companies throughout the world which may have relevance for your portfolio holdings and on industry developments more generally. If you own an innovative US-listed biotech company addressing a particular therapeutic area, you will need to know what the competition is doing and, most particularly in the biotech sector, you will have to be thinking about who could potentially acquire your portfolio shareholdings in the fullness of time, as this is often a key 'exit' for biotech investments.

Even though you might be based in the USA, you will need to have a clear idea of what Swiss-listed Novartis is up to, and Takeda in Japan, and GSK and AstraZeneca in the UK and Sweden, Novo Nordisk in Denmark, and myriad other such companies all over the world. One of the easiest and most logical ways of doing this is to be more or less ruthless about only looking at companies that are above a certain value or size. With the 'market data' tools available nowadays, this is relatively easy to do (with a Bloomberg terminal, for example). You might decide to cover a universe of companies above $10 billion in value, or $5 billion perhaps. The higher you set that bar, the easier your life will become given you will have several hundred fewer companies you need to follow and think about.

Applying the Pareto principle, you will probably be able to get 80 per cent or even more of the result you need in your job from setting that bar pretty high – and, by extension, end up with a manageable role, rather than be utterly swamped with far too much information and an impossibly large number of companies, products and assets to think about. Of course, this means that you will have neither the time nor the inclination to look at any companies outside of that core coverage universe, no matter how 'exciting' they may be. You just can't run your job like that and stay sane.

In addition to that, in buoyant years for the biotech market, at any one time you may also have two dozen or more investment bank salespeople calling you every week wanting you to give consideration to their deals – whether that be raising more money for existing companies (given how often biotech companies need to raise new money), some of which will be in your portfolio already, or for entirely new companies which they are looking to bring to the market via an IPO. There were 78 US biotech IPOs in 2020 alone, for example. Giving consideration to those deals is a hugely demanding endeavour in terms of the time and effort required to form a view on whether to invest or not.

There are several thousand listed biotech and healthcare companies globally and tens of thousands more private (unlisted) companies. To have any chance whatsoever of succeeding in their role, a given biotech analyst or fund manager will therefore have to triage the universe of companies to which they give their attention very aggressively indeed.

With few exceptions, they just can't be looking at small companies from outside of the USA which come with the additional headaches of being thousands of miles away, in a very different time zone, and of currency risk and complicated jurisdictional considerations such as tax to think about too.

In my former role, I used to joke (somewhat bitterly) about the fact that I could have suggested to a US biotech investor that one of the companies I represented had effectively cured all cancers or had developed a technology to turn lead into gold perhaps, and the response would still have been: 'I'm really sorry but I just don't have the bandwidth to offer a call with management right now.'

2 The 'too much capital' problem

As if the 'human resource problem' wasn't bad enough, the other problem is that specialist biotech investors in the USA tend to be running very large pots of money. The largest US biotech funds can be running several billion dollars. When you have that much capital to get to work you are very much forced to 'hunt the whale' and focus on larger businesses.

Imagine you are tasked with looking after $3 billion of investment capital and don't want to own more than, say, 50 companies – given the work required to keep on top of that number of companies. Mathematically, this implies that your 'average position size' will need to be around $60 million. The reality is that such funds have a number of much bigger positions at the top of their portfolio in bigger companies and then a larger number of smaller ones, but the general point still stands. It is simply not workable for you to even think about investing in a small European or Australian company which might trade only $500,000 worth of stock each day or, in many cases, even significantly less than that.

If you were to make an exception and support the IPO of such a company with even a 'half-unit' (of $30 million, for example), this would imply that you might end up owning far too high a percentage of that company than you would ordinarily be comfortable with and that you would likely need many months to sell out of it if you changed your mind, even in a good market.

The other point to make, which goes back again to the 'human resource problem', is that you may well need to do nearly as much work on the smaller company as you might have to do on a $10 billion company to get sufficiently comfortable to make an investment.

In terms of the ROI (return on investment) of your time and effort, it just isn't worth your while to give consideration to the smaller company, particularly when you factor in those additional points about distance, time zone, currency risk and other complications inherent in owning foreign companies. Such investors would much rather make a $100 million investment in a $10 billion value company based in the USA rather than invest that same $100 million in a much smaller company, even if the $10 billion company has already gone up a great deal in terms of share price and the smaller company may have that exciting upside ahead of it.

The market is 'inefficient'

This reality introduces an interesting question, for students of financial markets at least. For essentially all of the time that I have been looking at the biotech sector, it has been more or less clear that a company based in the UK or Europe with a given set of biotech assets is invariably valued a long way below what a company in the USA with similar assets tends to be valued at. This valuation discrepancy is often even more pronounced for companies listed in Australia if only because 'capital depth' in Australia is even more challenged than in the UK and Europe.

This *should* be what professional investors would describe as an 'arbitrage opportunity' – something that many investment funds are looking for explicitly day by day. All other things being equal, if a company with certain assets and intellectual property is valued at, say, $1 billion trading on the Nasdaq stock exchange in the USA, and a company listed in London with very similar assets addressing the same or similar therapeutic end-markets is valued at only, say, $200 million, then surely US and other investors might take notice of that sizeable valuation discrepancy, see it as a significant opportunity and look to buy shares in the UK company – thus driving the price up and 'arbing' away the clear valuation differential?

Financial market theory would tell you that this is precisely what should happen – in particular something called the efficient market hypothesis, or EMH. The idea is that market participants the world over should be looking at all available inputs and data-sources in the consideration of how companies are valued by a given stock market. Any clear valuation discrepancy should lead to buyers of the cheap assets and, sellers of the expensive assets until everything is priced 'correctly' or 'efficiently', all other things being equal.

This is a nice theory in the ivory towers of finance academics. Back in the real world, however, and certainly in the market for small biotech companies, this theory simply doesn't hold for the reasons outlined above. The global market for biotech science has tended to be thoroughly *in*efficient.

Market failure

Taken together, these factors go some way to explaining the extremely counterintuitive reality that a company valued at, say, $200 million can very often be significantly *less* attractive to the global investment community than one valued at, say, $1 billion, even if it may have similar assets and commercial prospects in the long run.

That is to say that the investment community as a whole can often be more comfortable buying the shares of a company which may have a price *five times higher* than another company, quite often even if it is possible to argue that those two companies are doing fairly similar things and the valuation discrepancy may not be justified by 'fundamentals'.

For a century or more, traditional economic and financial market theory previously suggested that the lower the price of something, the more attractive it is to potential buyers. In the real world, however, the opposite can sometimes be the case – for long periods of time at least, as a given market remains in some kind of significant 'disequilibrium' and can suffer from some kind of 'market failure' for a very specific set of reasons.

To be fair, this reality is reasonably well understood by economists and financial market participants nowadays, some of them at least. Since 1978, no fewer than five Nobel Prizes in Economic Sciences have been awarded for work done on 'Behavioural Economics', all five of which sought to explain precisely the sort of market failure and disequilibrium I am referencing here, to one extent or another at least. For example, Richard Thaler, according to a Nobel Prize press release, 'won the 2017 Nobel Prize for 'incorporating psychologically realistic assumptions into analyses of economic decision-making. By exploring the consequences of *limited rationality* [...] he has shown how these human traits systematically affect individual decisions as well as market outcomes.'

In plain English, big US investors would rather pay a higher price for biotech assets based in the USA because it makes sense for them to do so personally on many levels. This is why huge valuation differentials can and do exist which can have far less to do with the fundamentals of what a given company is actually doing commercially, and far more to do with the structure of the investment industry, structural challenges to the efficient functioning of stock markets and increasingly well-understood aspects of behavioural finance.

Many smaller biotech companies outside of the USA struggle to attract the interest of *generalist* domestic smaller company investors in their home markets and are too small, illiquid and far away to be interesting or even visible to large *specialist* healthcare investors in the USA or elsewhere in the world who might better understand them as an investment.

Given the above, there is a highly problematic 'funding gap' faced by companies outside of the USA, particularly when it comes to those which are big enough to need to start thinking about being listed on a stock market.

It gets worse...

These challenges around funding and investment result in a cascade of other problems for the development of the industry outside of the USA. Because it is so hard to raise money and build large businesses; it is impossible to build the same quality of self-reinforcing infrastructural 'eco-system' that has arisen in places like Boston and San Francisco. There is insufficient critical mass, and there are too few economies of scale across the industry as a result.

Companies which are starved for capital in the UK, Europe, Asia or Australia are competing globally for scientific and executive talent, for example. With less funding overall, they can't afford to offer the same salaries or perks available in the USA. It has been estimated that a young scientist with a biotech PhD can command about $50,000 a year in the UK as against more than $70,000 in the USA, for example. Further up the ladder, senior executives with the range of scientific and financial skills required to lead a biotech company are rare enough to begin with. With valuations so much higher in the USA and with sophisticated options packages and a more favourable tax regime, there is a real 'brain-drain' problem, with many talented British, Asian or European executives choosing to take jobs in the USA rather than stay at home and help build their domestic industries.

Similarly, because the industry overall is so much smaller, there is a less developed infrastructure of the highly specialized professionals that biotech companies need to help them succeed, whether that be lawyers, consultants, accountants and tax advisers or investment bankers. This problem is particularly pronounced across Europe when you consider that each individual country has its own individual rules and regulations, making the market for the right advice, whether legal, financial or regulatory, highly fragmented and complex.

With too little capital available to the industry historically, there are also real challenges around the supply of something as pedestrian yet mission critical as lab space. At the time of writing, British property consultants Bidwells have estimated that there is demand for more than 1 million square feet of lab space just in Cambridge in the UK, yet only about 10,000 feet currently available. It is thoroughly depressing that, in a university city which has contributed more Nobel Prizes in the relevant science than pretty much anywhere else on the planet, scientists cannot find enough lab space.

These challenges are circular and self-reinforcing. Because there is a lack of investment capital, biotech companies can really struggle to grow and succeed commercially. Because they struggle to grow and succeed commercially, in places like the UK, Europe and Australia they are seen as extremely risky by investors, governments and financial regulators alike, which is why there is that lack of investment capital in the first place.

British and European investors in particular have tended to be conservative when it comes to stock market investment generally, with the City of London in particular, focusing on 'established companies that produce dividends, not start-ups that lose money', as UK journalist Emma Duncan explained in a 2023 *Times* article. None of this is helped by financial regulation and the tax regime in places like the UK and Europe. British pension funds, for example, confront particularly draconian rules when it comes to giving consideration to making investments in riskier assets. This has prevented as much as £1.5 trillion of pension assets in the UK from supporting the sector. This is capital which is available in other parts of the world, particularly in the USA.

It is perhaps worth mentioning in passing that these challenges aren't just a problem for the biotech industry. These same points and themes go a long way towards explaining why the US tech industry has created several trillion more of value than anything yet to come out of the UK or Europe, for example.

These challenges have been bad enough in the past that many of those biotech companies outside of the USA that *do* manage to attract the attention of US VC investors have chosen to ignore their own domestic stock market and look to float on the US Nasdaq stock exchange. In the four years between 2018 and 2021, 22 European biotech companies floated in America. There are several more Australian businesses that have chosen to do the same, and of 14 British biotech companies which listed on a stock market in that same time period, no fewer than 11 of them listed on Nasdaq in the USA. This means that, according to Duncan, 'a firm's centre of gravity shifts towards America. In the end [...] the jobs, profits and tax revenue from which Britain might benefit go to America instead.' It is also one of the reasons why, as Chris Nave, director of one of Australia's few specialist healthcare investors, Brandon Capital, explains: 'Australia consistently ranks as one of the top nations for medical research, but one of the worst for bringing those discoveries to market.'

It is perhaps worth mentioning that these overseas companies who have managed to pull off a US IPO are a pretty select club of the lucky few. Listing in the US is enormously time consuming, difficult and expensive.

Even though a US listing can give such companies a better chance of cultivating the interest of large US specialist investors as against their peers who are listed back home, they are still thousands of miles and several time zones away from that investing audience. Overseas management teams may also not have the same 'cultural capital' as their US peers who are likely to be far more effectively networked in the USA having attended the same universities as many of the investment professionals so key to their potential success, or even quite possibly living locally to them and socializing with them more or less regularly.

Being listed in the USA is also extremely hard work for overseas management teams given how much time they have to spend in the USA post-IPO. An investor roadshow across a number of American cities several times a year is a particularly gruelling prospect for a European management team, let alone an Australian one as you might imagine. Taken together, a US stock market listing is certainly no panacea, particularly when the market for biotech as a whole is less buoyant than in the past, as many of these companies have endured in recent years.

All too often, such companies attempting to straddle their home market and a stock market listing in the USA end up moving their entire operations to America or giving up the idea of being independent and selling out. This is another reason why there are relatively few large biotech companies outside of the USA.

Gordon Sanghera is the CEO of one of the UK's few larger, stock market listed biotech companies, Oxford Nanopore. As he puts it: 'A company is offered £500mn to be acquired by a US company and everyone cheers. But why don't we say "no" and go and get the top talent and say, "Let's make it a £5bn company."' Or as Alexis Dormandy, former head of Oxford Science Enterprises has put it about UK start-ups: 'More often than not they are acquired for hundreds of millions by companies in the US and Asia. The founders pat themselves on the back [...] and we collectively celebrate our success. But that's exactly the moment we need to be asking how we build billion-dollar companies, and after that, ten-billion-dollar companies.'

All of these structural points about the investment industry, government regulation, tax regimes and insufficient critical mass to develop supportive ecosystems have made funding for smaller, innovative companies outside of the USA extremely challenging. This is slowing the progress we could be making scientifically and, by extension, human progress and aggregate wealth creation too.

Yet there are plenty of reasons to hope that many of these challenges will unwind or improve in the reasonably near future. Before we come on to look at those reasons, it is worth highlighting another factor which is retarding progress: the role of the press and, most particularly, the tendency of both the mainstream press and social media to have an extremely strong inherent bias to the negative.

Negative press bias

In my time working as an adviser to smaller biotech companies, all of the challenges faced by those companies were compounded by a strange phenomenon I encountered time and time again: a seeming total lack of interest in such companies from the mainstream press.

This was yet another extraordinarily frustrating and really rather odd reality confronted by those of us working with such companies and by the companies themselves. For some reason, even businesses doing really quite fantastic and consequential things in areas as important as cancer care, for example, simply never got the oxygen of any meaningful press coverage.

I suppose you can understand that there may be an element of career risk for a journalist being seen to promote work being done by small, 'risky' and often unprofitable companies. Being overly effusive about such a company one year and seeing it go out of business the next would certainly not be a good look for a journalist mindful of their credibility as a market commentator.

This being the case, there is unquestionably a chicken-and-egg problem here. These companies face an incredibly challenging funding environment for all the reasons I have given. This problem is compounded still further by the fact there are vanishingly few mainstream journalists willing to write anything about what they are doing, even when what they are doing is potentially really quite significant and even when such companies and their advisers are willing to take the time to personally contact those journalists and highlight that work. Some decent press coverage would be really helpful for their potential trajectory and help them get bigger and less 'risky', with significant second-order benefits for patient care, human progress and for the economy more generally.

The other challenge when it comes to press coverage, however, is more fundamental. It is a well-established empirical fact that journalism has a very strong built-in bias towards the negative and sensational and away from the positive and/or pedestrian. The vast majority of 'the news' that you are presented with every day is fundamentally negative. Extraordinarily so. We are all painfully aware that social media is no different.

Of most relevance for our purposes is the impact this reality can have on scientific progress and medical best practice. Study after study shows that bad news garners substantially more attention than good. This means that the mainstream press tends to pay far more attention to war, famine, murder, disease and political corruption than to more uplifting news stories about transformational new technologies.

The reason that our news is the way it is has to do with how human psychology functions at a very fundamental level, how this impacts all of us as the audience for 'news' and how journalists, editors and media companies conduct themselves as a result.

The problem is that we human beings have some extraordinarily problematic 'software' in our brains. We are hard-wired psychologically to focus on the extraordinary (and negative) to the exclusion of the unremarkable (and positive). Journalists and news editors know this, which is why they are particularly prone to the same phenomenon. This is what sells copy after all, precisely because we are all hard-wired this way. As the well-known media saying goes, 'If it bleeds, it leads.'

Possibly even worse than this, however, is the related problem that *false* (negative) news can often travel faster than the truth as a result. This is particularly problematic in the realm of science and scientific progress and a real challenge for biotech companies as I'm defining them.

In her book *Hacking the Code of Life: How Gene Editing Will Rewrite Our Futures*, British biologist Dr Nessa Carey, gives a number of examples of where this structural problem has cast a long shadow over the development of things as potentially important as gene editing, genetically modified (GM) crops and vaccines – all crucial technologies, several of which we will look at in more detail later in the book. Regarding GM foods she cites the example of a scientist named Arpad Pusztai, who, in 1998, claimed on TV that 'rats which had been fed genetically modified (GM) potatoes were stunted and their immune systems were suppressed'. His claims had not been peer reviewed, Carey points out, and yet: 'The fall-out was immediate [...] and played a huge part in the furore around GM foods and their possible impact on human health. A review of the science by representatives of The Royal Society concluded that the data did not support the conclusion that had been drawn. But the damage was done.'

Such media-channelled damage can be long-lasting, as Carey shows again and again. Another very high-profile case concerns a debate over the use of the ubiquitous measles, mumps and rubella (MMR) vaccine which continues to rage today. In 1998 the British journal *The Lancet* published poor-quality, highly compromised research by Andrew Wakefield claiming an association between the development of autism and the MMR vaccine. While it has been unequivocally proven there is no such connection and that vaccination against childhood diseases has in all likelihood been the most beneficial health innovation of the last century, Wakefield's 'research' continues to fuel online denunciations of vaccinations as a cause of autism.

The other broad point Carey makes is that 'retracted papers don't disappear'. This, combined with all those innate and imperfect cognitive biases we have looked at can mean that '"exciting" bad science may linger in the collective psyche much longer than the good boring stuff that corrects it'.

Carey's is by no means the only voice highlighting these problems. This is a theme that comes up time and time again in scientific literature and in books on popular science. It is also a topic of conversation which cropped up more or less frequently in my meetings with biotech company management teams and professional healthcare investors as well as at industry conferences.

I think it is worth highlighting just how high the stakes can be here. The damage done by this imperfect functioning of human psychology and how it intersects with the press and the effective dissemination of good information can be hugely impactful for the value of companies working in the area and for their ability to raise money and make progress as a result. Bad science, even when challenged and corrected and even for many years following that challenge and correction, can be similarly impactful. It can wipe hundreds of millions, sometimes even billions, off the value of companies working to deliver really important technological and scientific progress and can retard scientific and commercial progress by years.

Even more important, as a result it can mean far more human suffering in terms of illness and death than could otherwise have been the case. A number of the most exciting areas of potential innovation in the next few years are at particular risk here, not least given how many of them raise complicated ethical, economic and even political questions.

Medical best practice travels poorly

A key related challenge here is the fact that even in our supposedly incredibly advanced and networked world, medical best practice can take many years, sometimes even decades, to achieve ubiquitous adoption across the world – even in the supposedly 'modern' or 'developed' world. This is partly to do with the point I'm making around poor media coverage of so much of the great work being done, but it also happens as a result of several other intractable structural challenges to do with human nature, institutional inertia and health economics.

I think it is fair to say that when many people think about healthcare systems, in the developed world at least, they assume that they are basically the same, or at least reasonably similar wherever you might live. If you have a serious health problem or need an operation of some kind, surely your prospects and your likely experience and health outcome should be broadly the same whether you live in London, Edinburgh, Paris, Stockholm, Geneva, New York, Boston, San Francisco, Tokyo or Sydney, or, indeed, any other similar city of the rich developed world? Shouldn't this also be the case if you live in the developing world but are sufficiently wealthy to secure 'Western' standards of treatment?

This was certainly my working assumption for the whole of my adult life, until I started working with biotech and healthcare companies and, more specifically, came across the work of the American surgeon, writer and public health researcher Dr Atul Gawande, who in late 2014 gave that year's BBC Reith Lectures. Dr Gawande was ranked as the fifth most important thinker in the world by *Time* magazine in 2010. Among other things, he is a professor at Harvard University and has the kind of astonishing résumé a position like that so often implies. Not only is he a qualified medical doctor and surgeon but, interestingly, also a graduate of the University of Oxford's prestigious Philosophy, Politics and Economics (PPE) course too.

Perhaps of most relevance to the subject at hand is Dr Gawande's 2009 book, *The Checklist Manifesto: How to Get Things Right.* The core concept of the book revolves around the notion that checklists have facilitated some of the most challenging human endeavours, ranging from piloting airplanes to constructing remarkably intricate skyscrapers. By extending this principle to the intricate and diverse realm of surgery, Dr. Gawande developed a concise checklist that, within ninety seconds, significantly decreased fatalities and complications by over a third. He subsequently established the non-profit organization Lifebox and authored the World Health Organization's 'Surgical Safety Checklist,' which has demonstrated a reduction of complications and fatalities stemming from unsafe surgery by as much as 40 percent *globally*. These sorts of numbers have been achieved whether in advanced developed-world hospitals or out in the field in the developing world, which is the particular focus of Lifebox. Early research in Scotland demonstrated that more than 9,000 lives were saved, just in Scotland, in four years of using the list. By 2019, many years later, more comprehensive research had established a 36.6 per cent reduction in postoperative mortality across the whole of Scotland following a nationwide implementation.

Dr Gawande's 2014 Reith Lectures showed just how extremely imperfectly distributed medical best practice is globally and sought to explain why, often with fairly moving anecdotes. He began his first lecture with a story about how his own son, Walker, had a near-death experience in 1995 when he was only 11 days old. Based in Boston, Dr Gawande makes the point that his son was extremely fortunate to benefit from a very particular set of skills and an understanding of his son's condition that just happened to be present in the emergency room of his nearest hospital. At the medical facility, the team effectively identified the issue and prescribed a medication known as Prostaglandin E2. Dr. Gawande highlights that this drug had only been discovered approximately ten years prior to his son's birth. Additionally, his son's life was saved through cardiac surgery, which involved replacing his malformed aorta and repairing the existing heart defects. His son made a full recovery and went on to lead a perfectly normal life. Crucially, Dr Gawande then goes on to add about his experience with his son that day:

> In the ICU next to him was a child from Maine, about two hundred miles away, who had virtually the same diagnosis that Walker had. And when he was diagnosed, it was too long before the problem was recognised, transportation could be arranged and he could get that drug to give him back that open circulation, and the result was that poor child with the same condition my son had in that very next bed to us had gone into complete liver and kidney failure, and his only chance while we were waiting there was that he was waiting for an organ transplant to give him some chance at a future that was going to be very different from what my son had gotten to have.

Dr Gawande's family are originally from a rural village in Maharashtra, India. His broad point in reciting this story was to stress that, while it was perhaps obvious that had his son been born in India, his prospects in this situation would have been bleak, it was perhaps less obvious that the same could be said of a child from only two hundred or so miles up the road from Boston. This is a reality throughout the world. Whether or not you have access to the best possible medical outcome is incredibly capricious wherever you live, even today.

This was one of several experiences that led Dr Gawande to the holistic idea of a checklist as a mechanism to improve patient outcomes very significantly by reducing the risk of errors and accidents and even potentially to achieve healthcare outcomes that previously may have been considered impossible. He has been honing the approach, seeking to demonstrate its efficacy with robust evidence and advocating for its universal adoption ever since. One insight he and his team had very early on was that there were other areas of human endeavour where checklists have been universally adopted – in construction and the aviation industry for example as already mentioned above.

When looking to design a checklist for surgical procedures back in Boston, for example, his team invited the lead safety engineer from Boeing to help them. Over many decades, the aviation industry has gradually evolved away from a position of being one of the most dangerous ways to travel to being statistically by far the safest. In 1929 there were 51 fatal air accidents. This would equate to more than 7,000 per year nowadays, based on the current numbers flying. Can you imagine living in a world where 7,000 planes crashed every year? Even just between 1970 and 2019, fatalities per *trillion* 'Revenue Passenger Kilometres' (RPKs) decreased just over 80-fold from 3,218 in 1970 to 40 in 2019.

There are obviously many reasons for this breath-taking improvement, a crucial one of which is the standardized use of checklists across the aviation industry, by pilots and technicians alike. The same is true in the construction industry and at places like NASA, semi-conductor manufacturers and, indeed, in many areas of human endeavour where there is very significant complexity and where getting things wrong can cost billions and/or lives.

On arrival in the operating theatre as an observer, that Boeing engineer was astonished to discover that the surgical team didn't seem to have any codified plan or systematic process in place to ensure everything was done correctly. Even worse, in the aviation industry, there are only two or possibly three people in a cockpit at any one time. In the clinical setting there can be many dozen, thus growing the number of potential failure points exponentially.

Over some way more than a decade now, the checklist approach to the clinical setting has consistently demonstrated an ability to improve patient outcomes, reduce injury and mortality and also to save significant amounts of money for healthcare systems.

These facts notwithstanding, it has taken Dr Gawande and his team many years to get such systems and processes reasonably comprehensively, if imperfectly, adopted across the developed world, and adoption in the developing world remains patchy even today.

Something which should have been a relatively 'easy sell' based on many years of real-world evidence and, one might argue, on common sense too, was seemingly no such thing. As Lifebox has reported, factors which account for slow and low adoption have included 'surgeons who resent the implication that they may make dangerous mistakes, lax enforcement by hospital administrators, and the powerlessness of nurses in some cultures'.

Three months after the checklist programme was originally implemented in eight cities around the world back in 2007, the team surveyed the surgical staff to ask what they thought of it. That survey found that 20 per cent or more disliked it. Many viewed it as yet another time-consuming annoyance and source of yet more 'unnecessary' paperwork. Powerful senior staff in particular were perhaps understandably resentful of having to tick their way through such 'baby-steps' after a multi-decade career as a leading surgeon.

Interestingly, however, when the survey asked those same medical professionals 'If you're having an operation, would you want the team to use the checklist?', 94 per cent of them did. You might argue that the rather brilliant design of that questionnaire had more than a little to do with the success of the programme in the years that followed, compounded by the tangible results which were demonstrated once the data was in.

'Satisficing' surgeons

Hot on the heels of Dr Gawande's Reith Lectures in 2014, I attended my first investor lunch presentation given by a company whose lead product was a relatively new 'regenerative bioscaffold' for cardiovascular repairs. Key to their presentation was a great deal of personal testimony from leading cardio-thoracic surgeons in various parts of the world suggesting that the product was demonstrably achieving better patient outcomes than various other existing approaches or 'standards of care' that had been used for many years.

One of the key challenges levied at the company by the various professional investors around the table concerned the company's likely ability to drive wide adoption of the product, something they would need to do in order to succeed commercially. Notwithstanding the fact that this company had compelling data showing that its approach delivered better patient outcomes, and a number of the world's leading clinicians willing to support that position, people who knew about such things were still entirely cynical about the prospects for the company's technology.

It became clear to me that this was an ever-present concern. Here was a real-world example of precisely the sorts of challenges presented by Dr Gawande. It was the first of many such experiences in the eight years since then. Even as recently as only a few months ago I was asking an acquaintance of mine who is a leading surgeon in a certain therapeutic area about a company with an innovative technology addressing precisely that area. That person had not heard of the company, even though it has been promoting its technology to relevant clinicians for some time.

It is perhaps worth taking a moment to interrogate why this is so frequently the case. It has to do with a natural institutional inertia and what economists might describe as 'satisficing' behaviour on the part of medical professionals. Satisficing means making a choice or decision based on the first available workable option – a behaviour that is certainly understandable and probably forgivable given the context, even notwithstanding what a negative impact it has on the ability of healthcare systems to make optimal progress.

Imagine you have trained for as long as 20 years to become a surgeon. It might then take you another decade or more, perhaps even longer than that, to become a leading practitioner of a particularly complicated, life-saving surgical procedure or set of procedures in your particular area of specialism. Even at the cutting edge of global best practice, your approach may still have a relatively high mortality rate for all sorts of reasons. Then you go to one of the many international conferences which surgeons attend each year and see one of your peers, perhaps from the other side of the world, present a new approach which they are confident delivers better patient outcomes including lower mortality rates.

There are several entirely understandable structural reasons why it may then take literally years before you and your healthcare system are willing to adopt and implement that approach. First, doing so could very likely require a significant change in working practices with all that this implies for (re)training staff who, as we have seen, have already spent decades learning how to do it 'their' way. This is understandably a very big ask.

As economists would have it, there is a very significant 'sunk cost' problem. To overcome that reality, the new technique will have to show extremely compelling data-based evidence that it really is better and really does deliver better patient outcomes. Gathering that evidence can often take several years. Such things are also necessarily seldom black and white. Although evidence-based medical research will seek to use 'hard' data, there is still real nuance and complexity in such things and debate may rage for some time about whether a new procedure really is unquestionably 'better'.

This problem will be compounded further by the reality of differing health economics across the world. A new approach which is extremely expensive, either because it works with expensive drugs or requires a large capital investment in new equipment, may not be an option for healthcare systems with fewer resources at their disposal than others. As a 2013 report commissioned by the UK Royal College of Surgeons put it:

> The fruits of research are of little value if they are poorly implemented. Discovery only matters if it reaches and benefits patients. Spreading innovation in surgery is an attractive principle, but it can be difficult to achieve in practice. The diffusion of surgical innovation has posed particular challenges, from evidence, to training, to capacity. For example, the absence of appropriate evidence underpinning new techniques can inhibit investment in skills or infrastructure and limit clinical and patient demand for the innovation.

Economists use the expression 'satisficing' to explain the behaviour of market-participants in various settings, most particularly company management teams and employees. As Nobel prize-winning economist Herbert Simon said in his 1978 Nobel speech: '[D]ecision makers can satisfice either by finding optimum solutions for a simplified world, or by finding satisfactory solutions for a more realistic world.' In more common parlance, we might describe this as the well-known concept of 'if it ain't broke, don't fix it'. This is perhaps an entirely understandable approach for incredibly busy and resource-constrained professionals confronting very significant complexity in medical systems throughout the world and where decisions can make the difference between life and death.

What this does mean, however, is that medical practice can take a very long time indeed to spread around the world, with all that this implies in terms of our ability to improve our healthcare systems and ensure that the benefits of scientific progress are more equally distributed globally.

This has much to do with the role played by our press. In the global market for (good) ideas the press plays a vital role in disseminating information about medical advances, breakthroughs, and best practice as it develops to key stakeholders within society. Press coverage can help build trust and credibility in new medical technologies and play a crucial role in shaping healthcare policies and guidelines. Media attention can drive public pressure and the political will to prioritize the adoption and funding of those technologies.

With a lack of coverage or with excessively negative or inaccurate coverage, the general public, policymakers and even healthcare professionals can remain unaware of important medical or technological breakthroughs. Even worse, they can be misled and misinformed by information which is fundamentally inaccurate, leading to unwarranted fear and scepticism from the general public and poor policy.

In summary

As we have seen, there are some very fundamental challenges to scientific progress in many parts of the world. Historically, there has been a clear funding gap outside of the USA caused by the inefficient functioning of stock markets and the investment industry. This makes life hard for large numbers of biotech companies even in modern developed countries with a multi-decade track record of developing fantastic early-stage science and with world-leading institutions of learning.

These financial challenges are compounded by how little press coverage so many potentially exciting companies receive and, more generally, by an overwhelming negative bias in our mainstream media which gives far too little attention to exciting and potentially world-changing technologies which need investment capital.

This inherent negative press bias can also mean that bad science gets as much or even more attention than good. These factors can slow the adoption of medical best practice across the world and negatively impact the implementation of effective government policy too.

All too often, these challenges can strangle amazing potential treatments and technologies in the cradle. There is a vast amount of good science being done that isn't seeing the light of day commercially which is a tragedy for patient outcomes, potential wealth creation and human progress overall. This makes no sense given that the science elsewhere in the world is every bit as good as science in the USA that ends up being successfully commercialized.

In the next chapter, we will look at some of the many reasons these challenges will most likely be overcome.

3
The future is bright

The challenges I have outlined in Chapter 2 may seem intractable, the systems too entrenched. Crucially, however, there are many reasons to hope that things may improve. The direction of travel is encouraging. Many of the technological and societal advances which are coming will go some way to improving matters, and the march of human progress will alleviate some of these problems and solve others.

There are already signs that large global investors, governments, and academic and learning institutions alike are waking up to the challenges around funding for the sector. The crisis of undervalued biotech companies outside of the USA is increasingly being seen as an opportunity by smart investors with deep pockets, even if in the past they have been reluctant to look beyond their domestic market.

Look at the UK, for example. Oxford Science Enterprises was founded in 2015 to help commercialize science coming out of Oxford University. As the company puts it: 'We partner with scientific luminaries to translate novel research into new medicines that improve patients' lives.'

In the time since Oxford Science Enterprises was founded, it has raised over £800 million of its own capital to support companies coming out of Oxford. Even more encouraging, however, is that it has also helped attract a great deal more additional capital for its portfolio companies from more than two dozen of the world's smartest healthcare investors. The number of science-led companies coming out of the University of Oxford has increased from an average of four per year in 2015 to over 20 in 2021, and investment has lifted from an average of £125 million per year from 2011 to 2015 to more than £600 million between 2016 and 2021.

Likewise, lab space in Cambridge may be very tight (as we saw in Chapter 2), but there is a great deal in development more generally. In London, the specialist investor Kadans Science Partners is looking to invest £500 million in a new 823,000-square-foot, 23-storey 'vertical' life sciences campus. Elsewhere in the UK, specialist developer Bruntwood SciTech, in collaboration with Legal & General, one of the UK's largest investment companies, has 5 million square feet of property in development in 11 specialist science clusters throughout the UK in big cities like Birmingham, Manchester, Liverpool and Leeds.

The root of such changes lies in the nature of investment itself. As Warren Buffett's great mentor, Ben Graham, famously said: 'In the short run, the stock market is a voting machine. Yet, in the long run, it is a weighing machine.' What he meant is that significant short- or medium-term deviation from any kind of 'fundamental' value which can happen as a result of the sort of factors we have looked at in Chapter 2, will ultimately 'come out in the wash', in the (very) long run at least. If companies are unjustifiably undervalued, based on what they are actually doing commercially, eventually this will resolve. Put another way, as traders like to say, 'The cure for low prices is low prices.'

There are signs that this is what is happening. The challenges around funding for companies in many parts of the world are attracting increasing attention from the press, from governments and from investors alike.

Tangible commercial delivery

Most important of all here will be the fact that many companies working in the biotech industry outside of the USA could be on the cusp of delivering a tangible commercial outcome, even despite how hard the funding environment has been for them historically. This is why the real promise inherent in the sector is more likely in front of us than behind us.

The value of a given biotech company should increase as it successfully navigates clinical and commercial milestones, all other things being equal, and particularly if it can reach profitability. Many of the companies which have been so aggressively held back in recent years have nevertheless managed to make pretty significant progress through their clinical and/or commercial trajectory – even if they have been forced to do so 'running on fumes' financially.

I first met many of the dozens of companies I have worked with over the last several years when they were at a very early stage of their development. Plenty of these companies are now much closer to creating later-stage value. The global investment community that may have ignored them when they were many years away from delivering anything tangible commercially will be increasingly willing to ascribe value to businesses that can deliver later-stage clinical assets or who have secured regulatory approval for a medical device, for example. It won't take much for many of these companies to move from being valued in the tens or low hundreds of millions to being valued at more like several hundred million.

Crucially, this is often a key inflection point for so many of these companies. It will mean that they finally appear on the radar screens of large global investors who need to write those big cheques. There are numerous large investors who have already run the slide rule over many of these companies and who could be supportive with large quantities of investment capital, but only once those companies have delivered a positive outcome at their next clinical inflection point, and only once there is sufficient liquidity available for those investors to get enough of their capital to work.

You might describe this as a kind of 'investor ladder'. There is a series of 'gearing events', or clearly discernible step-changes, in the investor audience a given company can address as they continue to make progress and grow. This can mean that such companies can increase in value fairly quickly when they finally 'get there'.

As frustrating as the experience of the last few years has been for so many companies outside the USA, many of them are very close to attracting the 'oxygen' of global capital that was previously unwilling and or unable to give them serious consideration. For a company to succeed in navigating their way from a value in the tens or hundreds of millions to a value which 'begins with a b', one of the most important factors in their success is to focus on the right investor audience as they go and to migrate from one group to the next as effectively as possible as they deliver commercially.

Indexation

When it comes to companies which are already listed on a stock market, another factor that can drive their share prices up still further is something called 'indexation'. Throughout the world, when a company gets to a certain size, they will then be included in one of the main stock market indices. This would include the FTSE 250 Index in the UK and the ASX 300 in Australia, for example. The point is that there is far more global capital focused on those indices than there is giving consideration to non-indexed stocks.

A large number of investment firms tasked with running many billions of investment capital will only consider a universe of *indexed* stocks per their mandate – the rules which govern how they are permitted to invest. Often they are simply not allowed to invest in non-indexed stocks or, at the very least, permitted to deploy only a limited percentage of their capital in non-indexed companies.

There are also many funds which *must* take positions in a company once it enters a given index. These are tracker funds and exchange traded funds (ETFs) whose job it is to ensure they own all the constituent companies in a given stock market index on behalf of their underlying investors. Such funds account for many hundreds of billions of investment capital.

This is yet another example of a structural factor which can and does drive share prices with little regard for company fundamentals or the underlying *commercial* trajectory of those companies. It is one of the primary reasons that companies such as Apple, Amazon, Alphabet/Google, Facebook/Meta and Tesla have kept getting bigger and bigger for so many years, for example.

There are a good number of companies in the UK, Europe and Australia that are not that far away from getting to a size where they may enter a key local stock market index. This may create an ecosystem of 'mid-cap' biotech companies which didn't exist previously in many parts of the world outside of the USA.

The mere fact that there will finally be a crop of success stories valued in the billions will shine a light on the sector, to put it mildly. Large investment banks will take far more of an interest as there will be enough of an economic incentive for them to do so.

The largest investment banks such as JP Morgan, Goldman Sachs or Morgan Stanley basically don't 'do' smaller companies. It simply isn't economic for them with their cost base and business models. They need to be turning over hundreds of millions worth of stock and doing multibillion dollar mega-deals to pay for their 2-million-square-foot offices and costly headcount. In the vernacular of one pouty supermodel, these folks 'don't get out of bed' for a 100 million dollar company. Actually, depending on the animal spirits of what the stock market is up to at any particular moment in time, they probably don't get out of bed for even a 1 billion dollar company nowadays.

If there are a decent number of larger companies in future, there will be far more support from those investment banks. All other things being equal, this will mean that, in turn, there is far more interest from the investment community, too.

Success begets success: a trickle could very well turn into a flood. The result could be hugely positive for value creation and, far more importantly, for patient outcomes and scientific progress overall.

'One times forward earnings'

In terms of commercial delivery, it is probably worth stressing that eventually many of these sorts of companies will deliver tens or even hundreds of millions of *profit*, assuming they manage to navigate their clinical and commercial glide-path successfully.

Of course, there is plenty of risk that companies in this sector fail clinical trials or fail to get regulatory approval for a medical device perhaps. When this happens, share prices can fall a long way. If a company has only one asset, there is a chance it could become worthless, one of the reasons that generalist smaller company investors have avoided such companies in the past.

That said, there is a fair bit of work available on *average* rates of success at the various clinical inflection points for biotech companies working on drug treatments. Across many thousands of clinical trial results over more than a decade, it has been established that there is roughly a 30 per cent chance of a Phase II asset making it through to Phase III (see Chapter 2 for an explanation of clinical phases). There is then about a 52 per cent chance of a Phase III asset being approved. It is also possible to lift those averages by focusing on disease areas and approaches to drug development which have consistently shown a statistically higher likelihood of success in the past, and this reality is likely strengthening as innovative new technologies are applied to drug development.

There are plenty of companies out there, and outside the USA in particular, which are valued at tens or low hundreds of millions of pounds, dollars, euros or Australian dollars which could generate roughly that much in profit in the fullness of time, assuming they make it to market. What this means, big picture, is that someone willing to look at these sorts of companies might be able to construct a portfolio of companies where a decent number of them may effectively be on 'one times forward earnings'. My feeling is that there may be investors out there who view this as a compelling proposition at some point in the near future.

In the fullness of time, some of these companies will be making tens or even hundreds of millions of profit. It is likely that the market will then value them at more than a billion, and they may then enter a stock market index, too. Alternatively, such companies will be acquired by large pharmaceutical companies which need their intellectual property to bolster their product pipeline.

As small companies become medium-sized companies, not only do they become potentially more attractive to global investment capital, but they also crop up on the radar screens of the business development professionals at large pharmaceutical companies who are tasked with looking for acquisition targets. This is another reason why the 'efficient market hypothesis' so often doesn't hold for small biotech companies. Business development professionals are incredibly busy and have thousands of companies and technologies to look at all over the world. As a result, they too often 'don't get out of bed' for any business that is below a certain size, no matter how interesting or valuable their intellectual property might be in the long run.

As such businesses hit key clinical milestones and become just a little bit bigger, however, they can finally get to a size where people with very deep pockets may take an interest. Such folk have a great deal of capital. Ernst & Young estimated that the global pharmaceutical industry had nearly $1.4 trillion available to acquire such companies by the end of 2023, as we saw in Chapter 1. A great deal of that money will need to be put to work in the next few years as large pharmaceutical companies need to buy in the innovative technologies which will enable them to continue to deliver growth in revenues and profits in the years ahead.

Another key point is that *profitable* companies are likely to be of more interest to *generalist* investors, all other things being equal. Those biotech companies that survive long enough to reach profitability will be far more likely to attract the interest of generalist investors who can then evaluate them as a potential investment based on the conventional valuation metrics we mentioned earlier, such as profit multiples.

This is why those companies which *can* navigate these very particular structural challenges can be some of the best-performing companies in their respective stock markets in time. A handful of companies have managed this in the past; many more should do so in the future.

As an increasing number of companies deliver commercially and scientifically, there is a chance that what was once a 'toxic cocktail' of circumstances will morph into an upward spiralling 'virtuous circle'. This could mean that, finally, many of these companies, and the sector as a whole outside the USA, may enjoy some real momentum after a very difficult few years.

Front-page news

The other factor which could help in this respect is that some of the technology that will be delivered and the treatments that result could well be front-page news and finally attract a great deal more press coverage than has been the case in the past.

As I said in the introduction, there is a great deal of 'magic' going on out there and science fact is increasingly looking like science fiction. A company which delivers an effective 'cure' for a certain kind of cancer or genetic disease will be a good deal more likely to achieve meaningful press coverage as compared to when it was at a relatively early stage and several years away from delivering that cure. Similarly, an innovative medical device which might save many thousands of lives and strip billions of cost out of healthcare systems may eventually be of interest to journalists once it is on the market as against when it was an exciting but unproven idea in a laboratory.

There will be increasing interest in the sector overall as an understanding of the pace of scientific progress becomes more widespread and more and more 'revolutionary treatments' make the headlines, whether for cancer, diabetes, cystic fibrosis, Parkinson's disease, Alzheimer's, multiple sclerosis or so much else besides.

It will be the combination of that commercial delivery from many of these companies and a handful of very high-profile success stories that result which will finally attract the attention of a much bigger audience and bring significantly more capital into biotech companies that sit outside of the USA, too.

Exponentials and consilience

Many of the challenges faced by the sector will be solved more or less inexorably by the inevitable mathematics of exponentials.

In their excellent book *Abundance: The Future is Better Than You Think* (2020), Peter Diamandis and Steven Kotler explain that the human brain is simply not built to think in terms of exponentials. To illustrate the point, they compare taking 30 'linear' steps (1, 2, 3, 4 and so on) versus 30 'exponential' steps (1, 2, 4, 8, 16 and so on). We can all estimate with reasonable accuracy how far 30 steps might take us. Most of us, however, can't fathom the idea that 30 exponential steps would take you around planet Earth 26 times.

When only 1 per cent of power generation comes from clean energy, for example, it may seem intuitive that it could 'never' account for 100 per cent of power generation, but the rate at which these things develop is exponential not linear. In terms of the overall percentage of our energy production, in many parts of the world clean power generation does not grow and has not grown at a 1, 2, 3, 4 annual rate but rather at a rate of 1, 2, 4, 8, 16 and so on. In many places, it now 'only' needs to double once or twice more, year to year, to comprise 100 per cent of energy production. The same can happen across a whole raft of similar challenges.

This last point about needing only one or two final time periods to deliver a 100 per cent outcome is key, and another reason why we can often find it hard to conceptualize the pace of exponential progress. An idea which is often used to illustrate this concept is that of 'the lily pond', described by John Becher, a US executive named one of the world's most influential CMOs by *Forbes Magazine*, in his *Manage by Walking Around* blog. He tells us to picture a large pond with only one lily pad. The lily pad multiplies exponentially, covering the entire pond in three years. Every month, the number of lily pads doubles, resulting in full coverage of the pond in 36 months. If asked when the pond would be half-filled with lily pads, most people might assume it would take 18 months, halfway to 36. But the reality is that it would take 35 months. Just before the pond is fully covered, it's halfway there because it doubles the next month. Understanding this concept is straightforward, but our brains don't naturally grasp exponential growth.

He goes on to explain:

> The pond is 1/64 full in month 30, only 6 months before it is completely full. This is the problem with anything subject to exponential growth. It's deceiving. 30 months into a 36 month phenomena is a long time (2.5 years!) *but there are hardly any results*. Because the pond is barely full, many people won't believe the trend is real. Even those closest to the phenomena might not understand how big it is going to be.

Importantly, Becher uses a real-world example – the adoption of the smartphone. As he explains, in the early 1980s McKinsey & Company advised the US telecommunications giant AT&T, 'not to enter the mobile telephone business, predicting there would be fewer than one million cellular phones in use by 2000'. The reality was that by the end of 2000 there were 100 million mobile phone subscriptions. McKinsey's prediction was off by

99 per cent. 'Like many people before and since, the McKinsey consultants suffered from linear thinking.' Today nearly 7 billion people have a smartphone. This problem of linear thinking in an exponential world goes a long way to explaining why we are highly likely to be overly pessimistic about the world and, more generally, about the biotech industry and what it is likely to deliver for all of our benefit.

This is particularly the case when we consider just how exponential present-day exponentials are. The very nature of exponentials is that they accelerate, and never more powerfully than at present.

It took only 11 years for social media to go from a standing start to reaching seven in ten Americans. Even *within* that trajectory it is worth considering the fact that the legacy social media platform Myspace reached 115 million users between 2003 and 2008 at which point it was eclipsed by Facebook which went on to peak at 2.5 billion users by the end of 2019. Fast-forward to today and OpenAI's chatbot, ChatGPT, has been the fastest-growing app in history, reaching 100 million monthly active users two months after launching. ChatGPT has achieved in a few weeks what took Myspace, the first social media 'phenomenon', more than five years. That is about a 30-fold increase in the speed of adoption of the technology.

A key dynamic at work driving this acceleration in exponentials is the increasingly complex *interaction* of many different technologies. It is the role played by the *convergence* of numerous interdependent and self-reinforcing technologies which means that science and progress can move faster than ever before.

Another idea related to convergence is that of 'consilience', a term coined in the late 1990s by the American biologist and Harvard professor Edward Wilson. In his book *Consilience: The Unity of Knowledge* (1998), he argued for the fundamental unity of different fields of knowledge, suggesting that all of our key academic disciplines, including the natural sciences, social sciences and humanities, are interconnected and should be studied together in order to achieve a more complete understanding of the world. In the time since Wilson developed the idea, technological development and convergence has given scientists the tools to become increasingly 'consilient'. While reductionism and hyper-specialism continue to be extremely important at the coalface of scientific progress, technology has facilitated increasing cross-disciplinary collaboration at the most fundamental level, and this is yet another pillar underpinning the acceleration of our exponentials.

In his book *Exponential: Order and Chaos in an Age of Accelerating Technology* (2021), Azeem Azhar identifies four key technologies which will be particularly important for exponential progress in the next few years: artificial intelligence (AI); robotics; synthetic biology such as gene editing and DNA synthesis which will enable us to engineer living systems with increasing precision; and digital fabrication, sometimes described as additive (rather than subtractive) manufacturing or, more generally, as 3D printing.

It is the increasing convergence of these technologies and the ability of scientists and researchers to be increasingly consilient across all of them which will continue to drive and accelerate scientific progress overall, with far reaching and fundamentally positive implications for our future. As Azhar puts it, 'We have entered a wholly new era of human society and economic organisation' – what he calls 'the Exponential Age'.

Biotechnologies have a paramount role to play in all of the above. Big picture, we might be said to be in what I tend to describe as 'a race between *Mad Max* and *Star Trek*': between a dystopian future where society is on the brink of collapse and where there is violent conflict for scarce resources, and a thrilling halcyon utopia where advanced technology seems to have provided our species with everything we need to live a good life and where the only remaining existential threats come from alien species rather than problems of our own making, such as lack of resources or environmental degradation. When we consider the phenomenal advancement of our species achieved in the last few centuries and the technological progress being made on so many fronts, there is a good chance we can deliver a *Star Trek* future.

Reasons for hope

At present the significant majority of people in the developed world believe that the world is getting worse. In a 2018 survey, only 6 per cent of American people surveyed thought the world was getting better. In the UK and Germany, the figure was only 4 per cent, and in Australia and France, only 3 per cent. There is plenty of evidence that the reason for this tends to be that the majority of people are, quite simply, fundamentally mistaken about the trajectory of most of the things that really matter.

As examples: most people believe global poverty is rising when, in fact, it has been declining for decades. Most people believe war, violence, terrorism and crime are on the rise, when the opposite is true, even with the terrible events unfolding in Ukraine and Israel and Gaza at the time of writing. People believe that climate change is increasing the threat of natural disasters when, in fact, there has been an astonishing decline in deaths *per capita* of the human population caused by earthquakes, floods and so forth over the last century or so.

The inherent negative press bias we looked at in Chapter 2 tends to lead media companies to give a great deal of attention to bad news generally and to stories about the bad actors in the world – industrial despoilment, corporate greed, 'bad pharma' and so on – but vanishingly little coverage to all the good – and there is a great deal of good out there. It's just that more of it needs to be written about and funded!

You might argue that much of our press is looking at a 'glass' that may be as much as 99 per cent 'full' but is choosing to focus the vast majority of its attention and coverage on the 1 per cent that is 'empty'. This fact, combined with our inherent psychological biases, means that most of us are doing the same. We might be forgiven for doing so given how much harder it is to find the positive news stories, but I would argue that this doesn't mean that we shouldn't at least try.

For the last six years, the Australian non-profit organization Future Crunch has been producing a list of '99 Good News Stories You Probably Didn't Hear About' for the year just gone. In 2022 its 99 stories were divided by category into: Human Rights, Conservation, Global Health, Decarbonization, Development and Animals. As Future Crunch puts it on the landing page of its excellent website: 'If we want to change the story of the human race in the 21st century, we have to change the stories we tell ourselves.' Or, as Albert Einstein put it a century or so prior to that: 'The world as we have created it is a process of our thinking. It can't be changed without changing our thinking.'

It is all too easy to find mainstream press articles and dramatic (and popular) Netflix documentaries which paint an incredibly depressing and alarmist picture of the state of the world and significantly more difficult to find stories which share the exceptional progress that is so often being made. Looking to find those stories is a worthwhile (and uplifting) endeavour, however. Let's look at – and potentially somewhat recalibrate – two major areas: the environment and healthcare.

The environment

In conservation and the environment, there has been a steady stream of commitments made to reversing the damage done globally, and there are countless innovative commercial enterprises, collaborative government organizations and non-profit organizations making a real difference. Too little of this makes it into the public consciousness. Not only is this bad for our 'idea' of the world, but it also makes it significantly harder for such entities to attract the capital they need.

In the last decade or so, private and public organizations have created several million square miles of Marine Protected Areas (MPAs) all over the world. These aim to conserve biodiversity by protecting endangered species and ecosystems, while enhancing sustainable fisheries, and are recognized by the UN as the key mechanism in addressing the impact of biodiversity loss and climate change on marine environments. Much of this has been driven by an organization called Our Ocean Conference which has secured more than $128 billion worth of funding for marine preservation initiatives since 2014.

In the last few months of 2021 alone, the Convention for the Protection of the Marine Environment of the North-East Atlantic (OSPAR Convention) created the North Atlantic Current and Evlanov Sea Basin MPA to protect a huge area of 1.5 million square kilometres off the coast of Ireland. On the other side of the world, the Eastern Tropical Pacific Marine Corridor MPA will protect another million or so square kilometres of Pacific waters off the coast of Ecuador, Colombia, Panama and Costa Rica, including the Galapagos Islands. Similar initiatives have already been in place for some years in places like Australia and New Zealand, the Seychelles and across a very large chunk of the South Atlantic.

The same has been happening back on land. Over and above the MPAs mentioned above, there are millions more square kilometres of protected conservation areas, national parks and reserves on land all over the world, and the direction of travel here has been encouraging for quite some time.

In 2021 Indonesia was able to announce that it had achieved four consecutive years of *declining* deforestation under the leadership of President Joko Widodo. Elsewhere, in 2021, India 'announced a 25 per cent increase in mangrove cover since the 1980s thanks to restoration efforts' over and above numerous other initiatives which have seen tiger populations double globally and lots else besides.

Even China, so long viewed as one of the very worst actors in the world when it comes to the environment, seems to be turning a corner. In 2021 the Chinese regime announced plans to build the world's largest national park system, a project they have been working on since 2015. Five new national parks will cover more than 230,000 square kilometres and protect nearly 30 per cent of their key wildlife species. There is clearly still a great deal of work to do, but China has at least been working on such things for many years now and you could argue that these sorts of developments would have seemed unimaginable only a generation or so ago. They are all too little reported, in the Western media at least.

The list of such examples is very long indeed. Those of a more cynical nature could certainly accuse me of cherry-picking over the last few pages, and there is almost certainly some merit in this position, not least given that there is a long way to go to solve these sorts of problems and reverse the damage done since the beginning of the Industrial Revolution. But this in no way changes the fundamental point I am making which is, quite simply, how little coverage these stories get.

Healthcare

In healthcare, and 'biotech' specifically, great strides are being made across the board. Consider cancer treatment, for example. Japan is consistently ranked as having one of the most advanced and effective healthcare systems in the world. Japan's ten-year cancer survival rate has been increasing steadily for years and now stands at nearly 60 per cent. This is a disease that was once said to be incurable, but the survival rate has steadily increased alongside medical advances. Prostate cancer in Japan now has the highest survival rate at 99.2 per cent, followed by female breast cancer at 87.5 per cent, colorectal cancer at 69.7 per cent and stomach cancer at 67.3 per cent. In the USA, according to cancer.org, 'The death rate from cancer [...] has continued to decline. From 1991 to 2018, the cancer death rate has fallen 31 per cent. This includes a 2.4 per cent decline from 2017 to 2018 – a new record for the largest one-year-drop in the cancer death rate.' There is a similar story in Europe, where, according to one recent study, 4.9 million cancer deaths have been avoided in the last three decades.

It may not be entirely naive to hope that effective cures for many cancers and many other diseases may not be too many years away given how fast the technology is moving. Such treatments could also soon be produced at a price that would make them available in the developing world. As we shall see, one of the main challenges for the most advanced drugs and technologies addressing conditions such as cancer at present is cost. The most cutting-edge drugs cost hundreds of thousands of dollars, sometimes even more than $1 million a dose to manufacture – but this is changing, and at an exponential pace. Later in the book we will look at companies which may be able to bring highly effective novel treatments to market for a fraction of the cost confronted by healthcare systems at the moment.

I have already suggested in Chapter 2 that a welcome 'silver lining' on the awful 'cloud' of COVID may be a step-change in the speed at which the world will be willing and able to approve new drug treatments.

Whether it is the quest for an effective cure for cancer or for the eradication of any number of other diseases, rolling back the environmental damage of the last century or more, the development of new technologies to improve agricultural productivity very significantly or to provide us all with cheap, abundant and clean energy, phenomenal progress is being made across the world.

Changing the story

Whether or not these various initiatives and others like them will ultimately solve our biggest problems, I think it is reasonably clear that it would help if they got a great deal more press coverage and if a great deal more of us were at least aware of them! As Future Crunch puts it: 'What if bad news wasn't the only news?' Instead, we find ourselves inundated with a constant barrage of apocalyptic predictions, so many of which are opinions masquerading as facts. Swedish activist Greta Thunberg, in her speech to the Houses of Parliament in 2019, said that humanity is about to 'set off an irreversible chain reaction beyond human control that will lead to the end of our civilization as we know it'. And the US politician Alexandria Ocasio-Cortez has said with extraordinary specificity: 'The world is going to end in 12 years if we don't address climate change.' These kinds of statements are then too often taken as gospel or, at the very least, broadcasted widely and uncritically. Yet plenty of sensible and well-credentialled scientists think they're entirely wrong and that such statements have no basis in the science.

Australian climate scientist Tom Wigley is a leading expert on climate change. He has been working on the science since the mid-1970s and in 1987 created MAGICC, one of the first climate models and one that remains in use today. He has been one of the key individuals to sound the alarm about human-made climate change over the last few decades. He is categorically no 'climate change denier', rather a scientist in the vanguard of identifying the problem. That fact notwithstanding, when asked his view on whether climate change actually 'threatens civilisation' he has answered: 'It really does bother me because it's wrong [...] All these young people have been misinformed. And partly it's Greta Thunberg's fault. Not deliberately. But she's wrong.'

I don't particularly want to wade into the debate on climate change. It is an incredibly complicated, controversial and nuanced subject. All I do wish to do is highlight the likelihood that, in common with many other such important topics, the debate tends to be incredibly skewed because the negative side of the argument *inherently* carries far more power than the positive, for all the reasons covered in this section.

It is so powerful, in fact, that we collectively forget just how wrong so many of the catastrophists have been for so many decades, even centuries. The American biologist Paul Ehrlich still appears in article after article, year after year, to argue that the end is nigh and gets plenty of attention when he does. This, despite the fact that he has a five-decade-plus track record of being consistently and spectacularly wrong. In his 1968 book *The Population Bomb* he predicted that hundreds of millions of people would die of starvation each year from the 1970s onwards, although he revised this to the 1980s in later editions. At his most extreme, in the 1970 Earth Day issue of *The Progressive*, he prophesied that in a single decade, between 1980 and 1989, some 4 billion people across the globe would perish.

Ehrlich was not the first person in history to have done this and he won't be the last. Thomas Malthus was making essentially the same arguments as long ago as 1798. In his book *An Essay on the Principle of Population*, he argued that: 'The power of population is indefinitely greater than the power in the earth to produce subsistence for man.'

Happily, Malthus, Ehrlich and many others like them have been wrong for quite literally centuries. One of the main oversights of pessimism is that it fails to take account of the fact that exponential problems can have exponential solutions. If you extrapolate the bad news and fail to do the same with how technology and biotech specifically might improve healthcare, agricultural productivity or energy efficiency, for example, you will only see a negative outcome.

This phenomenon of our progress outpacing our problems will more than likely continue and even accelerate because that is how exponentials work at the most fundamental level. As an example, it is instructive to consider the role biotechnology has to play in significantly improving how we produce food.

In his 2020 book, *Moo's Law: An Investor's Guide to the New Agrarian Revolution*, British investor Jim Mellon makes the case that 'global deforestation, and the extent of land use, can and will be reversed by the new production methods that are coming down the line. About 99 per cent of land currently used to rear animals could be released back for other use, making space for housing, rewilding or recreation.' It is perhaps worth dwelling for a moment on the extraordinarily positive impact our being able to rewild 99 per cent of agricultural land would have on our environment. Mellon's main point is that exponential technological development in the agriculture industry in the next few years will revolutionize how the world feeds itself, injecting massively more efficiency across the board and very significantly reducing the industry's impact on the environment.

The Netherlands – a tiny country – has such efficient agricultural productivity that it is one of the world's biggest exporters of agricultural products, second only to the USA. In 2021 it exported nearly 70 per cent as much as the USA did, despite having only about one-twentieth of the population and being only around 0.4 per cent of the size. At the other end of the scale, poor food systems and supply chains mean that almost 40 per cent of food produced in India goes to waste. The same is true in much of the rest of the developing world. These technology gaps can and will be narrowed, however. Technological innovation in areas such as hydroponics and the production of cultured meat will make food production vastly more efficient and local. This fact alone could go a long way to rolling back the pace of environmental destruction suffered in the last couple of centuries.

Similarly, when it comes to clean power generation, progress is geometric, not arithmetic. As long ago as 2019 the UK newspaper *The Guardian* ran a story by Jillian Ambrose entitled 'Renewable electricity overtakes fossil fuels in UK for first time'. That article made the point that fossil fuels had been 'four-fifths' of the UK's electricity production 'fewer than 10 years ago', yet by 2019 'coal-fired power was less than 1 per cent of all electricity generated'.

The percentage of Germany's overall energy production from renewables had gone from less than 5 per cent as recently as the year 2000 to more than 42 per cent in 2021. This phenomenon can and will continue to improve as new technologies improve efficiency. More recently, another British newspaper, *The Independent*, published a piece by Anthony Cuthbertson reporting on 'Researchers from Germany [who] have set a new world record in solar cell efficiency using the so-called "miracle material" perovskite'. The team from the Universities of Wuppertal, Cologne,

Potsdam and Tubingen developed a tandem solar cell using organic and perovskite materials – a combination they hope could one day replace the silicon-based technologies used in conventional solar cells.

In the realm of nuclear power, there are similarly exciting developments. Thorium is an element which is more plentiful than uranium and potentially safer and with less of a waste burden, too. China is not far away from testing its first thorium reactor and may be able to build its first commercially viable reactor by 2030. And then there is nuclear fusion (rather than fission), which is almost certainly still a long way from being commercially viable but could have an exciting role to play at some point this century.

'Five grand challenges'

The US NGO 'Breakthrough Energy' has identified 'five grand challenges' which together account for effectively 100 per cent of what is driving greenhouse gas emissions. Specifically, these are: electricity generation, transportation, manufacturing, buildings and agriculture. We have already seen how we might realistically aspire to revolutionize agricultural production and clean energy generation at an exponential rate in the next few decades. When it comes to manufacturing and transportation, there is another exciting phenomenon which could also deliver exponential results: 'collaborative consumption'.

As Rachel Botsman and Roo Rogers wrote in their prescient 2011 book *What's Mine Is Yours: How Collaborative Consumption is Changing the Way We Live*: 'The average lawn mower is used for four hours a year. The average power drill is used for only twenty minutes in its entire lifespan. The average car is unused for 22 hours a day, and even when it is being used there are normally three empty seats.' New companies and technologies are already addressing this last reality, most obviously players such as Uber and Tesla, though there are plenty more working to revolutionize how we use vehicles more generally. We are early in the development of this particular 'lily-pond' – but given the very nature of exponentials, in a world where many of us will have moved from car ownership to renting cars fractionally only when we need them, we may not be as many years away from using 90 per cent fewer cars than might appear to be conceivable at the time of writing.

Not even two hundred years ago, when the first railway services began in the UK, Victorian commentators were genuinely concerned that speeds as dangerously fast as 20 miles an hour might cause instant madness in railway passengers, possibly even death. Today many millions of people all over the world travel in great comfort on trains which travel at more than ten times that speed. The idea that we might reduce car ownership in the next century or so by as much as 90 percent probably seems as inconceivable to most of us as trains travelling at more than 200 miles an hour in places like China and Japan would have seemed to the vast majority of Victorians.

A similar rental model will likely continue to develop across a wide range of other products. A raft of new technologies will continue to facilitate such changes just as the smartphone has done with Uber, for example.

This sort of progress will also likely reduce our need to travel as much as we have done in the past and may continue to inject a good deal more efficiency into how and where we construct, use and power buildings. It seems reasonably clear that another silver lining from the COVID pandemic has been to show businesses all over the world that at least some of their work can be done remotely via video-conferencing apps such as Zoom, Microsoft Teams and Google Meet with all that this implies for reducing the need to travel and the need for office space (and perhaps even for fundamentally improving the human condition and our relationship with work too).

There were 43.5 per cent fewer flights in 2021 than there were in 2020. Obviously, this extreme fall in the number of flights was a function of lockdown policies all over the world, and the industry has bounced back from those lows, but it does seem likely that there may be a long-run structural change in how aviation functions. At the very least the airline industry will have to do their utmost to find more efficiencies wherever they can, because the economics will demand it. As a 2021 article on the post-COVID airline industry published by US management consultant McKinsey has put it: 'Unlike the 2008 global financial crisis, which was purely economic and weakened spending power, COVID-19 has changed consumer behaviour – and the airline sector – irrevocably.'

There are already airlines contemplating the idea of windowless passenger jets. Instead of actual, physical windows, planes would contain virtual windows where the passengers can enjoy images projected onto 'windows' from outside the plane using fibre-optic cameras. As Sir Tim Clark, the president of Emirates Airlines has put it, those images are 'so good, it's better than with the natural eye'. Removing windows could reduce the weight of an aircraft by up to 50 per cent, strengthen the fuselage, increase the speed and altitude at which a plane can fly, and significantly reduce fuel consumption and emissions as a result. This is another example of a relatively simple innovation, almost certainly within our power technologically, which could meaningfully reduce the climate impact of a key industry within the reasonably near future.

It remains to be seen where all such things settle, but I would argue that exponential development across a wide range of new technologies and considerable changes in how we farm, eat, live, work, manufacture, consume and travel will likely pay enormous dividends for greenhouse gas emissions and for the environment as a whole.

This may happen some way sooner than many of us fear but, at the very least thanks to the counterintuitive functioning of exponentials, it could all happen incredibly quickly several years from now when the 'lily-pond doubles' in that one final time period. This will still deliver the outcome we all need.

In summary

As we saw in Chapter 2, there are fundamental challenges to the development of biotech globally, outside the more favourable environment found in the USA. However, there is every reason to believe that many of these market failures will correct themselves naturally – and exponentially – over the coming years. Biotech companies and technologies look set to flourish internationally, and there are already early signs that the present uneven global investment playing field in this area is beginning to level out.

What's more (and despite the messages of doom and gloom with which we are daily bombarded by the media) the outlook for general human flourishing is far better than we are often led to believe. Massive strides are being made to address the admittedly daunting problems humanity faces (disease, poverty, environmental degradation) and exponential developments in (bio-)technology can and will play a central role.

In Part Two we will focus on one of these problem areas – health – and show how biotech is revolutionizing not only medicine but how we as individuals can manage our day-to-day wellbeing.

PART 2

Towards medicine 3.0

In this part of the book, we look at the history of medicine and at how modern medicine developed. We consider the extraordinarily positive impact the development of antimicrobials and vaccines, and a multi-century improvement in nutrition, hygiene and sanitation, has had on mortality and longevity for so many of us.

We then consider the explosive growth in a number of 'diseases of modernity' such as diabetes, obesity, epilepsy, inflammatory bowel disease, and a raft of other autoimmune diseases and mental health problems, the key role played by the microbial world in all of this, and how this has been related to the rise of modern medicine.

We show how biotech is best placed to deal with these challenges. If we are to combat these many 'diseases of modernity' – and lead better, healthier, longer lives more generally – we need to do a far better job throughout our lives and focus on genuine 'healthcare' rather than 'sick care'. We need to understand the complex interplay of our genome and biome and how this impacts the 'right' diet for each of us, for example.

Biotech can and will deliver the technologies and healthcare systems needed to revolutionize how we approach medicine and health at a fundamental level and move us towards what is being described as 'medicine 3.0' – a more holistic, personalized, data- and technology-driven approach which can keep us healthier and even happier throughout our lives.

4
A brief history of medicine

Before we look at some of the exciting new technologies being developed by the biotech industry in more detail, let's take a look at where medicine has come from to set the scene.

The Dark Ages of medicine

For much of human history, many societies believed that illness was a punishment inflicted by gods or 'spirits'. With no understanding of human anatomy, the role played by infectious microorganisms, nutrition or the functioning of our immune systems, societies all over the world tended to ascribe mystical or superstition-based causes for disease. Diagnosis and treatment, such as they existed, would invariably have been the preserve of a 'witch doctor' or 'shaman'-type figure whose power and position in society derived from their supposed ability to intercede directly with those gods or spirits on behalf of a patient or even an entire community, particularly one suffering from an epidemic of some kind perhaps.

Such characters and practices still exist today in many parts of the developing world and even in the developed world when you consider the enduring popularity of faith-based approaches to healing still practised by many of the world's leading religions and a range of other more 'alternative' traditions.

To be fair to such practices, many of them can have merit. Although the efficacy of such approaches to healing and health overall will necessarily be rather hit and miss, arguably the reason that so many similar practices developed almost universally across most human societies over thousands of years is because they often worked – to some extent at least.

Seen through the prism of modern science it is more or less obvious why this was the case. Witch doctors, shamans, faith healers or yogic mystics have often been able to 'cure' patients as a result of a combination of factors which have less to do with communing with the gods and rather more to do with a number of more practical habits which such treatments tended to have in common.

At the most basic level, those would include some kind of beneficial nutrition (and, just as importantly, hydration), enforced rest of some kind and, crucially, the often powerful placebo effect inherent in a given 'patient' having real belief in their ability to recover.

Wherever a sick person was in the world, whether in a village somewhere in rural Asia, the Brazilian rainforest, ancient Greece or somewhere in Africa, Australasia or a Pacific Island, common 'treatments' in many places would often suggest the ritual consumption of various herbs or of a particular 'sacred' fruit or other food. More often than not, those herbal remedies would also be delivered as a 'potion' or 'elixir' of some kind, increasing the chance that a given patient was properly hydrated in times and places where access to safe, fresh drinking water was rather more of challenge than it is for those of us alive today.

Other practices found across many cultures included enforced rest of some kind, whether that was 'baths, naps, and wine' in ancient Rome, a ritual recovery hut in an African, Asian or Latin American village, or the use of early versions of saunas and steam rooms in places as far-flung as Scandinavia, Turkey and Japan.

Perhaps most important of all, however, was the power of the placebo effect and of what is often described nowadays as the 'mind–body connection'. Both of these concepts have previously been poorly understood and even treated with considerable cynicism by serious medical scientists. An article for *EMBO Reports* by Vicki Brower summarizes the sea change in thinking here:

> In the past 30 years [...] research into the link between health and emotions, behaviour, social and economic status and personality has moved both research and treatment from the fringe of biomedical science into the mainstream. According to the mind–body or biopsychosocial paradigm, which supersedes the older biomedical model, there is no real division between mind and body because of networks of communication that exist between the brain and neurological, endocrine and immune systems.

Put another way, as Brower says: 'Mounting evidence for the role of the mind in disease and healing is leading to a greater acceptance of mind–body medicine.'

On the placebo effect, specifically, as recently as December 2021, Harvard Medical School published a piece entitled 'The power of the placebo effect', stating that:

> The idea that your brain can convince your body a fake treatment is the real thing — the so-called placebo effect — and thus stimulate healing has been around for millennia. Now science has found that under the right circumstances, a placebo can be just as effective as traditional treatments.

In a society where everyone around you believed that the methods employed by the village witch doctor or shaman were effective, this could often become a self-fulfilling prophecy, particularly when added to the other more pedestrian but similarly helpful elements of such a treatment listed above.

At the more esoteric end of such things, many cultures have also made use of powerful herb-based, natural psychedelics such as iboga in several African cultures, ayahuasca across the Amazon, peyote by indigenous North Americans, psilocybin (from 'magic' mushrooms) in many other parts of the world and other natural products with euphoriant and stimulant properties such as kava in the Pacific Isles and khat in much of the Arab world and East Africa.

In the past, conventional wisdom and the scientific establishment held that many of these substances were entirely dangerous and to be avoided. More recently, however, modern science has begun to look in detail at many such compounds. There is a great deal of serious clinical research being conducted with many of them and plenty of sensible clinicians and scientists who believe they could demonstrate real efficacy in the treatment of anxiety, depression, PTSD and drug and alcohol addiction. This is particularly exciting given how intractable and untreatable so many of these mental illnesses have proven to be to date. As *Time Magazine* has put it: 'For years, the field of mental health has been largely barren of meaningful treatment advances.'

Several natural compounds commonly used by indigenous tribes from all over the world seemingly may hold the power to change that, however. In 2019 US author Michael Pollan published a ground-breaking book on the subject: *How to Change Your Mind: What the New Science of Psychedelics Teaches Us about Consciousness, Dying, Addiction, Depression, and Transcendence*. As he says:

> [F]or most of the 1950s and early 1960s, many in the psychiatric establishment regarded LSD and psilocybin as wonder drugs for treating depression, anxiety, trauma, and addiction, among other ailments. As these drugs came to be associated with the 1960s counterculture, and as stories began to surface about bad trips and psychotic breaks, 'the exuberance surrounding these new drugs gave way to moral panic.' Now the pendulum is swinging back, and the interest in their usefulness as a tool to help treat a variety of psychiatric conditions is rapidly growing.

Many of these compounds are now the subject of serious clinical studies, and the market for such things is projected to reach several billion dollars per annum in the reasonably near future.

For much of human history, the reason that 'traditional' medicine could quite often work, to a certain extent at least, was simply a function of belief, nutrition, hydration, rest and, in many societies, potentially even an enlightened use of various natural psychedelics when it came to mental health too. The 'wisdom' inherent in traditional treatments, such as it existed, was invariably a function of centuries of observation and trial and error, and the communication of such methods by oral traditions passed down from generation to generation. Although this broad approach could often be better than nothing, it was quite clearly rather hit and miss.

Notwithstanding the positive outcomes such approaches were often able to achieve for many communities in many times and places, not everyone was so lucky. For every culture successfully applying natural remedies and methods which worked, there were plenty of others employing methods which did not or which could even make things significantly worse.

The historical and archaeological records provide us with plenty of examples of 'treatments' which might almost be considered amusing if they weren't so horrific, and many others that were simply entirely ineffective. The Romans liked to use stale urine for various medicinal purposes, for example, and, indeed, as a teeth whitener and even to launder their clothes.

Many of us will remember primary school history lessons which taught us about the widespread use of leeches and of bloodletting more generally, from antiquity until really quite recently. Bloodletting was often particularly severe and is thought to have caused or at least hastened the deaths of historical figures as consequential as George Washington and King Charles II as well as vast numbers of other poor souls across several centuries.

Another common medical practice found throughout much of human history is that of 'trepanation' or 'trephination' – that is to say, the process of drilling a hole into a patient's skull. This was astonishingly commonplace from the Neolithic period up until as recently as the Renaissance. The archaeological record has found skulls across the world showing the marks of trepanning, with an astonishing 5–10 per cent of all Neolithic skulls found having single or multiple skull openings showing use of the procedure.

For every witch doctor, shaman or other early medical practitioner prescribing an entirely sensible and often effective mixture of herbs, chicken soup, potions and rest, with worrying frequency there was another happily serving up compounds of varying toxicity, relieving you of several pints of blood, or attempting to drill a hole in your head.

Early societies were also broadly incapable of dealing with injury and had no understanding of pathogen-borne disease. Even relatively minor injuries could be fatal if a wound became seriously infected and someone unfortunate enough to receive a more serious injury in battle, or when out hunting perhaps, would have been unlikely to survive.

It wasn't until some way through the nineteenth century that we began to understand the role played by pathogens in disease transmission, and neither was there any understanding of vector-borne diseases such as malaria or dengue, caused by mosquitoes, ticks or flies for example, until really quite recently. This meant that early human societies took none of the precautions required to prevent such things from being hugely problematic. In fact, often our forebears did the precise opposite of what is obvious to us today, with terrible consequences. In many parts of the world, humans lived in close proximity to their livestock and, logically, as near as possible to supplies of fresh water, both of which habits exposed them to vector- and pathogen-borne disease.

More often than not, open defecation was the norm and sanitation was non-existent. Even as recently as the 1840s, handwashing was considered controversial. So controversial in fact that Ignaz Semmelweis, the young Hungarian obstetrician who, in 1847, became the first to suggest it might be a good idea, was ridiculed and eventually confined in an insane asylum where he was subjected to extreme 'treatments' such as being doused with cold water as well as beatings by guards, dying two weeks later. Happily, scientific debate is rather more tolerant nowadays and the downside of suggesting potentially innovative, life-changing ideas invariably rather less extreme.

Hardly surprising, then, that diseases when they came were invariably horrendously impactful, not least when combined with other factors such as crop failure caused by occurrences such as the 'little ice age' which could leave populations particularly vulnerable and malnourished. It is estimated that the Black Death (an outbreak of bubonic plague), for example, may have reduced the global population by between 30 per cent and 60 per cent in the fourteenth century.

In his 1651 book *Leviathan*, the English philosopher Thomas Hobbes famously described human life as 'solitary, poor, nasty, brutish and short'. There were many reasons for this, of course, but our ancestors' failure to understand disease and the lack of the most basic healthcare was clearly a very significant part of the problem. The development of the rational scientific method and of modern medicine as one of the happy consequences of that has been instrumental in lifting much of humanity out of that sorry state of affairs.

The development of 'modern' medicine

As concerns medicine specifically, the long road back from an existence that was 'nasty, brutish, poor, solitary and short' arguably began in earnest considerably more than two thousand years ago with the birth of the Greek physician Hippocrates of Kos. Hippocrates is often described as the 'father of modern medicine' and is believed to have been born around 460 BCE.

Earlier we looked at how many human societies had viewed disease and illness through the prism of superstition of one kind or another, believing that such things were likely to be a punishment inflicted by gods or 'spirits'. The journey away from such beliefs and to a modern, rational, science-based understanding of what makes us sick arguably began with Hippocrates and his followers. These important thinkers and healers are generally understood to have been some of the first characters in recorded history to explicitly move away from that belief in superstitious or mystical causes of disease to one based on observation and an attempt to codify some kind of *system* of treatment. They are also credited with fundamentally professionalizing medicine and establishing it as a distinct practice, promoting key ideas which are in use to this day such as diagnosis, prognosis, observation and detailed documentation such that progress could be recorded and shared with other physicians.

Many readers will be familiar with the Hippocratic Oath, one of the oldest binding documents in history that remains the basis for medical ethics to this day. The oath enshrines ideas such as patient privacy, non-maleficence, and the obligation to teach medicine to others and to the next generation. It is a testament to the powerful legacy of Hippocrates and his followers that some version or other of the oath is still sworn by graduating medical professionals all over the world, and many of its fundamental tenets can be found in law such that violating them can be a criminal offence in many countries.

As with so many other elements of Greek culture, Greek medical practice was assimilated and carried on by the Romans. The best known of the Roman medical practitioners, Dioscorides and Galen, lived in the first and second centuries CE, respectively. Dioscorides compiled the first pharmaceutical compendium, the *De Materia Medica*, while Galen developed the theory of humoral imbalance. A key point here is that their ideas would flourish for centuries, to the Renaissance and even to the Industrial Revolution, with astonishingly little progress made.

Although Hippocrates and the Romans that followed him were, to a certain extent, a reasonable forward step from what had gone before, medical practice was still pretty dreadful, and patient outcomes left a great deal to be desired. We have already seen how durable terrible ideas such as bloodletting and trephination were, and many centuries passed with plenty of other thoroughly imperfect Greek and Roman ideas dominating medical practice, in the Western world at least.

It is perhaps not overly controversial to suggest that the reason so little progress was made with medicine over those many centuries was because so little progress was made by humanity overall. The period between the fall of the Western Roman Empire during the fifth century up until the beginning of the Renaissance roughly one thousand years later is often described as the 'Dark Ages' both because of how miserable, barbaric and impoverished it was and also, as a function of that reality, because of how little was written or recorded and how little cultural progress was made as a result. Set against the stunning cultural achievements of the Greeks and Romans over the previous several centuries, much of the world did indeed 'go dark'.

It took a cascade of related and self-reinforcing cultural developments to change this and to herald a new era for human progress and, by extension, for medical practice – broadly, the combination of the Renaissance and the Agricultural, Financial and Industrial Revolutions which then ensued.

Although historical figures such as Copernicus and Galileo continued to have their work banned by the Catholic Church, one of the key features of the Renaissance was a gradual move away from the near total dominance of religious dogma and towards more enlightened ideas such as humanism and the application of the rational scientific method based on observation and reason. The Renaissance changed the way people thought and were permitted to think at a fundamental level.

This paid enormous dividends for human progress. The Renaissance period is probably most often associated with developments in art, literature, architecture and science. Just as consequential for human experience overall, however, were key *financial* innovations such as the development of double-entry bookkeeping (accounting), early bond and equity markets, and, as a result, modern banking not dissimilar to the system still in use today.

Little else could have happened without the development of accounting and banking and the ideas of credit, debt and equities because, quite simply, it could not have been funded. Such financial innovations meant that far more capital and resource could be concentrated on building things and solving things than had been the case throughout the Dark Ages.

The rational scientific method could not have developed in the way that it did without the economic surplus needed to free up and give patronage to a new class of thinkers, writers, scientists and, of course, medical professionals such as doctors and surgeons and to fund the facilities and equipment they required. Renaissance ideas, habits and institutions fired the gun on a vast number of interrelated and co-dependent exponential developments which finally broke humankind out of the Dark Ages and enabled us to make extraordinary progress as a species from then until now.

I think it is worth dwelling on just how many extraordinary innovations were required across society as a whole over the last two or three centuries to get us from where we were at the end of the Dark Ages to the immeasurably better place we are today, the extent to which we take so much of this for granted, and the fact that this means that we may fail to comprehend where we are likely going.

In his 2010 book *The Rational Optimist: How Prosperity Evolves*, British author Matt Ridley reminds us of just how instrumental the ideas of specialization and exchange have been for human development across the board. As he says:

> By exchanging, human beings discovered 'the division of labour', the specialisation of efforts and talents for mutual gain [...]. The more human beings diversified as consumers and specialised as producers, and the more they then exchanged, the better off they have been, are and will be.

Perhaps the most fundamental Renaissance-era development of all which underpinned this process like no other was the development of the printing press and of affordable paper. The ability to document ideas and then share them widely, and reasonably inexpensively as compared to what had previously been the case, was crucial for the development of everything else.

The extraordinary Renaissance-era explosion in absolutely everything – art, architecture, engineering, science and medicine, and the Financial, Agricultural and Industrial revolutions which followed – simply could not have happened were it not for humankind's newfound ability to record and share ideas like never before.

The printing press was also instrumental in the rise of a much larger and more complicated nation-state, and one which was now able to tax a far larger percentage of its citizens as a result. Income tax was only first introduced in the UK in 1798, for example.

This new, much wider tax base and growth in government infrastructure overall was instrumental in giving the state the means to build the 'great works' that then followed in the Victorian era, particularly running water and sanitation, which were also critical to the development of healthcare for obvious reasons.

Renaissance ideas and the application of the rational scientific method first facilitated an agricultural revolution. This was key to improving agricultural productivity across the modern world enough to then create the economic surplus required to support the Industrial Revolution which followed.

And it was the Industrial Revolution which was instrumental for the transformation of healthcare and medicine after so many centuries with essentially no forward progress. The Industrial Revolution provided the ideas, tools and concentration of capital required to build and equip hospitals and improve funding for other dedicated institutions such as university medical schools and laboratories, for example. Arguably most important in healthcare specifically, however, was the development of organic and then industrial chemistry, and of the pharmaceutical industry and the first pharmaceutical drugs as a result.

What did the pharmaceutical industry ever do for us?

I don't think it is overly controversial to suggest that in recent years the pharmaceutical industry has generally been viewed pretty negatively by a meaningful percentage of the population all over the world. In August 2019 a poll conducted in the USA by polling company Gallup, for example, found that 'The pharmaceutical industry is now the most poorly regarded industry in Americans' eyes, ranking last on a list of 25 industries that Gallup tests annually.' The Harvard School of Public Health noted that '58 per cent of Americans held negative views of the pharmaceutical industry while only 27 per cent held positive views of it' and that this was 'the lowest the industry has ever been ranked in the poll, which began in 2001'. That same Harvard article reported: 'Gallup noted that high drug prices, the opioid epidemic, and Big Pharma's big lobbying efforts are all factors that likely played into respondents' frustration with the industry.'

The poll was conducted before the COVID pandemic. Since the arrival of the pandemic, a number of subsequent polls have found that things may have improved somewhat, with 30–40 per cent of Americans saying that their view of the industry had improved as soon as April 2020 and 63 per cent saying that the 'pharmaceutical industry has a positive impact on society' in one poll conducted by Grayling in June 2021.

Many of these polls have relatively small sample sizes. That 2019 Gallup poll, for example, surveyed only 1,525 adults in the USA and the Grayling one only 3,000, reasonably thinly spread across the UK, USA, France, Germany, Russia and China. However, a larger IPSOS Mori poll conducted for the UK Association of the British Pharmaceutical Industry (ABPI) surveyed more than 8,000 British adults in March 2021 and found that '60 per cent of the public said their views of the industry had improved since the pandemic'. Interestingly that same poll established that 81 per cent of respondents admitted to knowing 'nothing' or 'just a little' about the sector.

This is particularly relevant to the biotech and pharmaceutical industry. The more people learn about things, often the more positive they feel about them, particularly if there is a fair bit of evidence that things are positive, as may seem entirely logical.

While perception of the industry overall may have improved on average, if these various poll findings are anything to go by, it seems clear that opinions have polarized pretty strongly. The smaller minority of people holding negative views are arguably more strongly negative than ever before when you consider the anti-vaccine movement and the meaningful percentage of Americans who believe that Bill Gates wanted to use coronavirus vaccination to implant microchips into US citizens as part of a global surveillance programme.

When you consider that reported deaths of Americans who have been vaccinated are running at 0.0025 per cent of the number of doses given at the time of writing, it seems that 'anti-vaxxers' are likely to be looking at a glass that is more than 99.9975 per cent full and choosing to focus on the 0.0025 per cent that is empty. In fact, it is quite some way worse than that given that clinicians are obliged to report serious health problems to the Vaccine Adverse Event Reporting System (VAERS) whether or not that health problem was caused by a vaccine. As the Centers for Disease Control and Prevention (CDC) explains: 'If a health problem [or death] is reported to VAERS, that doesn't mean that the vaccine caused the problem. It warns vaccine safety experts of potential problems that they may need to assess, and it alerts them to take further action, as needed.'

We do not know yet how many of those deaths may have been directly caused by a vaccine. What we do know is that, even if it was all of them, which is vanishingly unlikely, that is still 'only' one death in every 40,000 doses given. The real number is almost certainly a long way lower than that. It seems likely that a similar dynamic has been at play when it comes to people's previously negative perception of the pharmaceutical industry and of drugs overall.

Throwing the baby out...

There is no question that there have been numerous bad actors in the industry over the years, as evidenced by Ben Goldacre's undoubtedly eye-opening bestseller *Bad Pharma*. In his own words:

> Doctors and patients need good scientific evidence to make informed decisions. But instead, companies run bad trials on their own drugs, which distort and exaggerate the benefits by design. When these trials produce unflattering results, the data is simply buried. All of this is perfectly legal. In fact, even government regulators withhold vitally important data from the people who need it most.
>
> Doctors and patient groups have stood by too and failed to protect us. Instead, they take money and favours, in a world so fractured that medics and nurses are now educated by the drugs industry. The result: patients are harmed in huge numbers.

However, I would argue that to take these arguments at face value may be a case of throwing the baby out with the bathwater.

One of the main criticisms Goldacre himself levies at the industry, as we can see from the quotation above, is the fact that pharmaceutical companies and regulators regularly withhold or omit 'vitally important data' about their products. He gives plenty of evidence to that effect, but are he and others like him perhaps guilty of the same 'crime of omission' when they themselves focus only on the bad elements of the industry while failing to highlight the enormous good it has done for our species overall, big picture?

This was certainly the view of one reviewer, Dr Humphrey Rang of the British Pharmacological Society. Dr Rang acknowledges that *Bad Pharma* is 'a book with an important message that deserves to be taken seriously' but goes on to suggest that extrapolating a number of admittedly negative features of the pharmaceutical industry to suggest that 'medicine is broken' is a 'foolish remark'. As Rang says: 'The fact is that medicine is one of the most successful and valued enterprises in the civilised world and, as Goldacre himself acknowledges, it owes a great deal to medicines developed in the last few decades, despite the flaws in the process.'

I would agree. We might argue that a fairer assessment of the merits of the industry overall should give more serious consideration to the aggregate impact on the human condition that has come as a result of their products over the last century and more overall. Might we stop for a moment to consider just what an astonishingly positive difference pharmaceutical companies and their products have made to our lives in the round? Seen through that prism, things look rather better.

The other point that I believe is worth making is that the direction of travel here also seems encouraging. Earlier in the book I suggested that many corporate entities are getting steadily better as a simple function of the fact that so many of their various stakeholders care about such things, whether management, staff, customers or investors. Things are by no means perfect, but in criticizing the pharmaceutical industry, I think it is important to retain a sense of perspective about just how much good it has delivered and, crucially, also consider the extent to which things may be getting steadily better over time.

History and evidence

If improved knowledge may lead to improved perception, it is worth dwelling for a moment on what the development of the pharmaceutical industry has done for humanity overall. To appreciate how far we have come, imagine living in a world with no anaesthesia, no painkillers and, most particularly, no treatments for infectious disease. It is not an exaggeration to suggest that our ability to treat infectious disease and even, in some cases, to eradicate it altogether has been utterly transformational for the human experience in the last couple of centuries.

Two key technologies which have enabled us to do that have been the development of vaccines and of antimicrobials (antibiotics, antivirals, antifungals and so forth). Before the development of those two key technologies, hundreds of millions of our forebears were carried away by a vast array of diseases, many of which are preventable or curable today, thanks in large part to the development of the pharmaceutical industry.

To get a sense for just how extreme this reality was, as we have already seen, the Black Death, or bubonic plague, caused by the bacterium *Yersinia pestis*, may have reduced the entire global population by between 30 per cent and 60 per cent in the Middle Ages. Even as recently as 1900, according to the UCI School of Pharmacy & Pharmaceutical Sciences, 'one-third of all deaths in the U.S. were from three general causes that are rare today because they are preventable and/or treatable: pneumonia, tuberculosis, and diarrhoea'.

A huge focus for the last few years across the world has been on COVID-19 for obvious reasons. At the time of writing, it is estimated that about 6.9 million people have died of (or with) COVID. There are many who take issue with the accuracy of these statistics, given that they are very hard to collect and also given that the methodology in many countries counts people who died *with* a recent positive COVID test, even if they may not have actually died *from* the disease itself. Concerns have also been raised about the fact that there seem to have been far fewer deaths recorded from other respiratory conditions in the last few years.

But if we ignore those misgivings about the validity of the statistics, and take the number at face value, 6.9 million people equates to roughly 0.086 per cent of the world's population (of about 8 billion) who have died with or *of* COVID since the start of the pandemic. We might compare this with the largest pandemic of the twentieth century, the Great Influenza, or 'Spanish flu', of 1918–20. Estimates for the number of deaths globally during that pandemic vary from between just over 17 million and as high as 100 million. The global population at the time is estimated to have been about 1.8 billion, implying that the pandemic killed between 0.96 per cent and 5.5 per cent of the world's population – that is, between 12.3 and 70 times the deaths per capita caused by coronavirus.

As can be seen from these different experiences of a global pandemic, we have been getting rather better at dealing with such things over time, thanks in large part to the development of the pharmaceutical industry, antimicrobials and vaccines, as well as social, health and medical changes.

Antimicrobials

The US National Institutes of Health (NIH) defines antimicrobials as '[a] substance produced by one living organism that kills or inhibits the growth of another'. They are a broad class of therapeutic which includes antibiotics, antivirals, antifungals, antiseptics and biocides.

Professor Dame Sally Davies was formerly Chief Medical Officer of the UK (2010–19). In her important 2013 book *The Drugs Don't Work: A Global Threat* – a primer on the nature of infectious disease and the history of our ability to fight it by developing and using antimicrobials – she makes the point that antimicrobials, on average, extend people's lifespan by twenty years. Since penicillin was manufactured in 1943, humanity has survived remarkable surgeries and life-threatening infections. She underscores that infections have shaped the history of human disease and, at times, have influenced history itself. (Professor Davies made these points nearly a decade ago but how prescient they now seem!) She then goes on to chronicle the fall of infection in many parts of the world over the last century or more, providing numerous examples including the virtual disappearance, at least in the developed world, of infections such as puerperal fever and syphilis. Antimicrobials have enabled us to combat dozens of other infectious diseases. As a consequence, Europe and North America have witnessed an exceptional and unparalleled decrease in mortality for more than three generations.

Another key point that Davies made at the time was the extent to which such progress was not equally distributed across the world. While mortality rates and serious illness may have fallen and life expectancy increased in the modern, developed world, the same could not be said for much of the developing world. She points out that in 2011, out of around 55 million deaths globally, around 10 million, or roughly one-fifth of all deaths, were caused by infectious diseases. Although this was an astonishing improvement on historical rates, as we have seen, she goes on to point out that 9.5 million of those deaths occurred in low- and middle-income countries with 'only' half a million or so in the developed world. As she says, even as recently as 2011: '40 per cent of all deaths in low-income countries were a result of infectious diseases, compared to around 7 per cent in the UK and other high-income countries.'

Happily, things have improved a great deal even just since then. The total disease burden from communicable, maternal, neonatal and nutritional disease has fallen more than 22 per cent since 2011 and it had already fallen significantly for decades up to 2011. Exceptional progress has been made in the developing world in recent years across the piece, whether with malaria, polio, HIV, hepatitis C or any number of other diseases where we have very clearly been winning the battle for a long time. This steady improvement continues, and there are good reasons to expect it to accelerate thanks to the biotech industry.

Davies's particular focus in her book is on the role played by antimicrobials. The other key technology which has helped achieve such progress, of course, is that of vaccination.

Vaccines

Today the US Centers for Disease Control and Prevention (CDC) list vaccines for no fewer than 25 diseases that are more or less routinely used in the USA, not including the coronavirus vaccine. These include vaccines for: anthrax, chickenpox, dengue, diphtheria, influenza, hepatitis, human papillomavirus (HPV), measles, meningitis, mumps, pneumonia, polio, rabies, rubella, shingles, smallpox, tetanus, tuberculosis, typhoid, whooping cough and yellow fever.

It is worth taking a moment to dwell on what the above list implies for human health, happiness and longevity. Some of the diseases listed are more deadly than others, but it is quite clear that the development and widespread use of these vaccines and of antimicrobials has saved or, at the very least, extended hundreds of millions, most likely even billions, of lives in the last century or more, and completely transformed our lives and our relationship with disease.

In many cases, these technologies have eliminated or nearly eliminated previously fatal diseases, such as polio, tetanus, tuberculosis and malaria, or even eradicated them altogether as has been achieved with smallpox, for example. The difference between elimination and eradication is, broadly, that *elimination* is where we have stopped the spread of a disease but where continued efforts are required to maintain that position, whereas *eradication* is where no further intervention measures are required because the disease has been effectively wiped out entirely, most famously with smallpox which was deemed to have been eradicated by 1980. And there is much more to come. It is highly likely that the list of diseases we have managed to eradicate will grow in the years ahead as new technologies come on stream.

It took the pharmaceutical industry only about ten months to research, create and deliver a vaccine for COVID-19 (as against more like 45 years for polio). There are many reasons to expect and hope that we are on the cusp of seeing that kind of acceleration in progress across many other areas, including cancer, for example. As the cost of manufacturing and supplying such treatments comes down, we will also continue to see these innovations shared beyond just the rich, developed world. Going forwards, an increasing proportion of the global population will most likely be able to benefit from exponential developments in healthcare.

It's about modernity as much as medicine

Of course, the reason we are winning this battle the way that we are is not purely as a result of the biopharmaceutical industry. Plenty of things have influenced the decline in mortality including improving nutrition, hygiene and housing, but all of these improvements are relevant to my point about the march of modernity as a whole.

Our ability to make such extraordinary progress has been as a result of the convergence of many different yet vitally related factors. The development of financial institutions and markets has enabled us to pool sufficient capital and apply that capital to innovation across an astonishing range of crucial technologies. That innovation has given us powerful antimicrobials, vaccines and numerous other treatments, as we have seen, but it has also been critical for the development of everything else required for us to make such fantastic progress in the developed world and, increasingly, in the developing world, too.

As an example, our ability to get most vaccines to their end user requires 'cool-chain distribution'. As Hans and Ola Rosling put it in their 2018 book *Factfulness: Ten Reasons We're Wrong about the World – and Why Things are Better Than You Think*:

> Vaccines must be kept cold all the way from the factory to the arm of the child. They are shipped in refrigerated containers to harbours around the world, where they get loaded into refrigerated trucks. These trucks take them to local clinics, where they are stored in refrigerators. These logistic distribution paths are called cool chains. For cool chains to work, you need all the basic infrastructure for transport, electricity, education, and health care to be in place.

The tech industry has obviously been instrumental in the development of all of the above. Tech has had a crucial role to play in the development of more efficient refrigeration, better storage, modern harbour facilities, and in improving the efficiency of those trucks and of power generation more generally.

Tech has even had a role to play in providing the education and skills required to the personnel working in all of those fields and in healthcare, too, of course, now that medical professionals can be taught and trained remotely using the internet, for example. Tech applied to architecture and design has also been crucial for all of the above and for our ability to build ports, storage facilities, hospitals and health centres more cheaply and efficiently, particularly in the developing world. The fact that Moore's law has reduced the cost of processing power and data storage exponentially has also meant that we can analyse and improve every element of these complex supply chains like never before.

The quality and sheer scale of data we have access to nowadays and the analytical tools we have at our disposal to interrogate and make use of that data is yet another remarkable feature of modernity. At the time of writing, it is estimated that there are as many as 5 billion internet users worldwide. Throughout the course of writing this book, in common with those several other billion people, I have been able to look up almost any information I might need as it has occurred to me. I have been able to find academic papers, statistics, charts, tables, graphs and other source material in moments.

A few pages ago, I cited some of the COVID statistics and made the point that there are reservations about the quality of some of that data. I would argue that these reservations aside, the data that we *do* have access to is nothing short of remarkable. The fact that so many of us can access such an enormous set of aggregated and near-real-time COVID data pulling in from dozens of government healthcare systems and non-government organizations (NGOs) in milliseconds and for free is perhaps worth a moment's reflection. It may not be perfect, but it is astonishing nevertheless, and gives healthcare systems, practitioners, policymakers (and authors) the tools to make informed decisions as never before. It is certainly one of the (many) factors which we have to thank for COVID having been a fraction of the disaster that the Great Influenza was a century ago.

The same point can be made about any number of datasets concerning essentially every single area of human progress and innovation. Not that long ago, the most powerful computers in the world cost tens of millions of dollars and needed to be housed in vast, purpose-built industrial spaces. Even those computers were incapable of handling datasets of a size that a smartphone can handle easily today.

Today billions of us have those smartphones. Scientists and researchers can also enrol large numbers of smartphone users all over the world using sophisticated, intuitive and user-friendly apps to provide large quantities of rich and accurate data in real time for all sorts of research projects. We also have collaboration tools such as Skype, WhatsApp, Zoom, Microsoft

Teams, Google Meet and Slack. As Jennifer Doudna, winner of the 2020 Nobel Prize in Chemistry, said of her team: 'All told, we would be quite the international group: a French professor in Sweden, a Polish student in Austria, a German student, a Czech postdoc, and an American professor in Berkeley.' That team, based in several countries across no fewer than nine time zones, was able to conduct regular video calls, share their screens and post large and complicated files for analysis and consideration in real time. All of the above is unprecedented in human history as against every other generation of our species and, as can be seen, is resulting in work which is good enough to win Nobel Prizes and deserve to be described, without hyperbole, as being potentially world changing.

Today, scientists are able to collaborate and make progress like never before. Most areas of human endeavour nowadays, particularly at the coalface of innovation such as in the biotech industry, rely on scientists being able to share datasets and file sizes which would have been unthinkable even as recently as a decade ago. Far too many of us take this reality entirely for granted and it hasn't really sunk into our collective consciousness.

If the development of the printing press was key to our emergence from the Dark Ages, the quality, size and power of our networked world today is exponentially more consequential for our trajectory into the future.

Nutrition, hygiene and sanitation

More generally, improved nutrition, hygiene and sanitation have helped deliver the astonishing fall in rates of disease and improvement in mortality, longevity and health. Here again the tech and biotech industries have been instrumental in driving that progress. In many parts of the world, agriculture is exponentially more productive than it was even only a few decades ago and widespread cool-chain distribution is just as relevant to perishable foodstuffs as it is to vaccines, of course.

There has already been spectacular innovation in crop production, irrigation, water filtration, storage, distribution and also, crucially, in food testing and regulation. In many parts of the world our sustainable crop yields and food safety more generally would astonish the farmers and consumers of only a few generations ago.

Today, in most of the world we produce far more food and far safer food than would have been considered possible even relatively recently. The range and quality of foods available to many of us is also exponentially better than it was as recently as two generations ago. Notwithstanding high food inflation in much of the world at the time of writing, food still costs those of us in the developed world a fraction of what it cost our grandparents and great-grandparents as a percentage of our income.

There is also a fair bit of evidence that we are only just getting started here, in terms of agricultural efficiency and productivity and food safety and security. New (bio)technologies applied to agriculture and agricultural production and to distribution are going to improve such things still further. Such progress is likely to be exponential in nature with this being one area in particular where I believe that 'revolution' is more likely than 'evolution', and in my lifetime.

Crucially, these exciting innovations are also likely to be increasingly more evenly distributed across the world than was the case in the past. Happily, many of the new innovations and technologies in this area will go some way to levelling the playing field by reducing cost, significantly enhancing efficiency and making food production fundamentally more local.

Significantly improved standards of hygiene and sanitation have also been critical. We can thank the industrial chemistry, pharmaceutical, distribution and retail industries for inexpensive antiseptics and disinfectants, both of which have been critical for improved healthcare outcomes globally. As concerns sanitation specifically, in 2015 the United Nations identified 17 key development goals which were to be the focus for numerous governments, NGOs and corporations between 2015 and 2030. Number six of those goals was 'clean water and sanitation'.

Solid progress has been made in this area to date. In the developed world it is fair to say that most of us take running water and widespread and relatively advanced sewage and water recycling systems entirely for granted, given they have been around since before almost all of us were born. We probably also take for granted the fact that here, too, there is constant innovation as new technologies are discovered and adopted gradually. Such technologies would include advanced membrane filtration systems and reverse osmosis, UV irradiation, ion-exchange and electro-dialysis, 'bioaugmentation' of wastewater with sophisticated enzymes and even nanoparticle treatment where carbon nanotubes can be used to remove even the smallest pollutants from water.

These innovations help us improve how we use, recycle and purify water in the developed world. Crucially, many of them can also increasingly be deployed in the developing world and should go a long way to improving water security for many millions of people. To a certain extent, developing countries may even be able to 'leapfrog' the developed world here just as happened with telecom infrastructure, for example. As the US think tank the Center for Strategic and International Studies (CSIS) has put it:

> Leapfrogging occurs when a nation bypasses traditional stages of development to either jump directly to the latest technologies (stage-skipping) or explore an alternative path of technological development involving emerging technologies with new benefits and new opportunities (path-creating). Probably the most famous and regularly cited instance of stage-skipping is the mobile revolution, which put phones in the hands of millions of people while allowing developing nations to skip directly to mobile phones without the need to invest in landline infrastructure.

Might we hope that something similar can happen when it comes to water and sanitation systems in the developing world? While this is perhaps extraordinarily unglamorous subject matter, the humble toilet gives us an example of just this sort of thing.

Dr Doulaye Kone is the deputy director of Water, Sanitation and Hygiene at the Bill & Melinda Gates Foundation, having previously worked at the African Water Association (AfWA). He has the perhaps unenviable task of 'reinventing the toilet and reimagining fecal sludge management business as a sustainable utility service'. As he explains, nearly 250 years after the invention of the flush toilet, around 3.5 billion people – almost half the world's population – are forced to use unsafe sanitation facilities. Consequently, half a million children under the age of five die every year from diseases like typhoid, diarrhoea, and cholera. Additionally, many more people fall ill, leading to an estimated US$223 billion a year in health costs and lost productivity.

In 2011, this prompted the Bill & Melinda Gates Foundation to embark on a challenge to 'reinvent the toilet' and transform sanitation. The aim was to create a toilet that would eliminate dangerous pathogens, potentially convert waste into valuable resources for low-resource settings, and not require water, power, or traditional emptying to remain sanitary.

Bringing safe, affordable sanitation to the world would vastly improve the quality of life for the billions without it. It would also act as a 'super-vaccine' that would effectively end the spread of many deadly diseases, just as it already has in places where flush toilets are the norm.

Over the last decade, several hundred million dollars have been invested in this initiative by the Bill & Melinda Gates Foundation and numerous private companies and other entities. There is still a long road to travel, but dozens of companies are working on projects all over the world to reduce costs and make these innovations more widely available as well as affordable. This will pay huge dividends for global health, and probably sooner than many of us might believe.

The (bio)tech industry and innovative biotech companies have an enormous and potentially world-changing role to play here. Too few of us are even vaguely aware of just how many thousands of companies and projects there are out there working on such things, whether on therapeutics, agriculture, water and sanitation, power generation or numerous other related fields.

And again, to a certain extent, in many ways we are actually only just getting started. The rich, 'occidental' world has made extraordinary progress to date. Here, things are likely to improve still further. Even more exciting, perhaps, is that what will hopefully come next will be the steady spread of all these sorts of ideas and technologies to every part of the world.

Real wealth

Innovation has given us the specific technologies which have delivered this step-change in the human experience, whether antimicrobials, vaccines, improved agriculture, sanitation, power generation or lots else besides. More broadly, and very importantly, it is *financial* innovation that has also given us the real wealth required to pay for it all. It is the development of reasonably sophisticated banking and financial market institutions that has enabled us to pool enough capital to provide significant funding for innovation and 'great works' across the board.

Between 1996 and 2022, total global spending on research and development grew nearly 350 per cent in real terms for example, from around $555 billion to more than $2,478 billion. This steady and durable growth has happened notwithstanding sizeable financial market crashes in 1999/2000 (the 'dot-com' crash), 2007–9 (the 'Global Financial Crisis') and continues today, with an estimated $2.35 trillion invested in 2021 even when the COVID crisis was raging all around us.

Another way of tracking this kind of progress is to look at the growth in rates of patent registration globally (Figure 4.1 below). Between 1980 and 2020 patent applications globally grew more than 250 per cent from about 650,000 a year to more than 2.3 million in 2020. That year even saw significantly more patent applications versus 2019, despite the COVID crisis (though perhaps partly *because* of the response to the COVID crisis, of course).

It is perhaps worth highlighting that R&D spend and patent registration are key drivers of human progress and of the real wealth creation required to solve

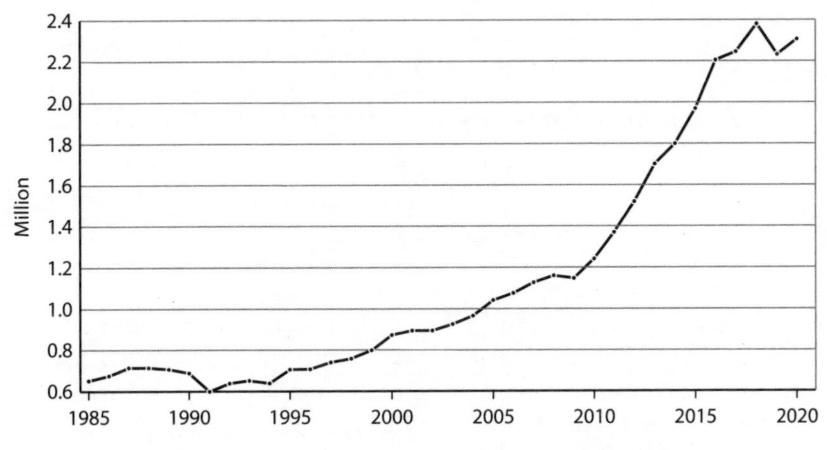

FIGURE 4.1 Global patent applications, 1985–2020

Source: World Intellectual Property Organization (WIPO), WIPO Report: Statistics on Worldwide Patent Activity, licenced by CC

our most intractable problems. An increasing percentage of all of the above is coming from the biotech industry, for the simple reason that this is where so much of the most exciting, potentially transformational cutting-edge science can be found and where the most powerful exponentials lie.

It is highly likely that both of these trends will continue into the future and may even accelerate given how exponentially more powerful the underlying technologies are and also given the geometric effect of 'convergence'. When you add biotech exponentials to tech, communication and network exponentials, things get even more exponential (as it were).

In summary

Modern medicine and the development of the pharmaceutical, biotech and tech industries have delivered a very significant improvement in the human experience in the last century or more as we have triumphed over infectious disease in particular. In the next chapter, we will look at the rise of the growing challenge of antimicrobial resistance and at the growth in a raft of various diseases of modernity – both of which phenomena are related to that triumph over infectious disease, as we shall see.

5

Antimicrobial resistance and the rise of 'modern plagues'

Let's return, for a moment, to Professor Dame Sally Davies and her book *The Drugs Don't Work*. The reason for the book's title, was because, having highlighted the profound impact of the development of antimicrobials to date, she used the rest of the book to highlight the alarming problem and 'global threat' of something called antimicrobial resistance (AMR). Professor Davies describes AMR as 'just as important and deadly as climate change and international terrorism'.

Her point is that, while antimicrobials may have added an average of 20 years to everyone's lives and given us so much else to date, including something as fundamental as the ability to practise surgery, the evidence of the last few decades is that these benefits are being steadily eroded as our antimicrobials have gradually begun to stop working. As the World Health Organization (WHO) explains:

> Antimicrobial Resistance (AMR) occurs when bacteria, viruses, fungi and parasites change over time and no longer respond to medicines making infections harder to treat and increasing the risk of disease spread, severe illness and death.

As a result of drug resistance, antibiotics and other antimicrobial medicines become ineffective and infections become increasingly difficult or impossible to treat.

Facing up to the crisis

In 2020, in reaction to this very significant problem, a consortium of more than 20 leading biopharmaceutical companies and industry associations contributed around $1 billion to launch an 'AMR Action Fund'. As they explain on their website:

> Our world faces a silent, slow-moving global threat that kills around 700,000 people each year due to the uncontrolled rise of superbugs resistant to antibiotics. These superbugs can affect anyone, of any age, in any country. AMR is a universal issue that impacts us all – we are all at risk.
>
> This looming global crisis has the potential to be as large or even larger than COVID-19 in terms of deaths and economic costs.

In their December 2014 'AMR Review' paper, produced at the request of the British government, Lord Jim O'Neil and his team estimated that the economic cost of AMR, if unsolved, could be as high as $100 trillion by 2050. For context, in January 2022 the International Monetary Fund (IMF) put the estimated total cost of COVID at $12.5 trillion. As the AMR Action Fund goes on to say:

> AMR undermines every aspect of modern medicine. We rely on the availability of effective antibiotics to be able to do everything from wisdom tooth extractions to organ transplantation to cancer chemotherapy. Superbugs could set medicine back to the 19th century, before the discovery of penicillin, when simple infections could kill.

It estimates that, unchecked, deaths caused by AMR could rise to more than 10 million per annum by 2050. Even this frightening number could underestimate the potential problem. As Sally Davies explains in her book, we must alter our conduct and acknowledge that antibiotics are a valuable asset that we must safeguard and utilize only when there is a clinically justified rationale. Failure to do so could lead to a scenario where 40 percent of the population die prematurely from infections that are untreatable.

Forty per cent of the population! A *Mad Max* outcome if ever there was one. But again, such exponential problems will very likely have exponential solutions.

One of the key reasons why AMR is such a threat has to do with how the commercial market for antibiotics functions at a fundamental level, or, more aptly, *doesn't* function. This is a result of a very specific problem which is unique to antibiotics as a treatment. According to the AMR Action Fund, there is no sustainable market for antibiotics. Creating antibiotics involves a lengthy, intricate, and uncertain process, with many candidates failing during development. After approval, new antibiotics are used cautiously to maintain effectiveness and slow the emergence of resistance. While this approach benefits public health, it doesn't encourage the substantial investment required to sustain a strong pipeline of antibiotics.

In plain English: no one can afford to invest billions to develop a product which will be used only 'sparingly'. This is why it has been almost 30 years since a new class of antibacterial has been discovered. There has been a little progress since Davies wrote her book in 2011, for example, but not much. Research published in July 2021 concluded that: 'The number of new antibiotics and their indications are not keeping up with resistance and the needs of the patients.'

At present, this very serious problem is another classic example of what economists describe as a 'market failure'. But there are two reasons for optimism in terms of our ability to solve the problem: the emergence of new technologies and a clear commitment from so many key players to find a solution.

First, the arrival of a number of completely new technologies: there are already biotech companies working on entirely new ways of killing microbes, which may mean that resistance is no longer such a problem. This could change the economics of the market as such new technologies would not need to be 'used sparingly'.

One such company is UK-based Destiny Pharma. Destiny has been developing its 'XF' platform for several years. According to its own website: 'The XF drug platform has a novel, ultra-rapid mechanism that reduces the chance of bacteria becoming resistant to its action.' Destiny has been able to show that its 'XF-73' product has a 'unique no/low resistance profile'. In March 2021 it published the results of its Phase IIb clinical study achieving a 99 per cent reduction in methicillin-resistant *Staphylococcus aureus*, more commonly known to us as MRSA – the cause of several hard-to-treat and potentially fatal infections.

To be clear this is 'only' a Phase II study (see Chapter 2). The company is still some years and quite a few million dollars of R&D (research and development) spend away from proving conclusively that this technology works, but their progress to date is certainly encouraging. Sadly, Destiny is also a classic example of a company suffering from many of the various challenges I highlighted in Chapter 2 in terms of access to capital and press coverage.

Another exciting technology which seems to come to us squarely from the realms of science fiction is that of 'nanobots'. Nanobots are microscopic 'robots' which are 'thousands of times thinner than a human hair and able to cruise around a body and repair a bone or heal an illness'. This kind of technology has appeared in fiction for decades, whether in the *James Bond*, *Star Trek* and *Terminator* franchises or in best-selling novels by the likes of Neal Stephenson, Robert Ludlum and Michael Crichton (of *Jurassic Park* fame).

The idea to develop such technologies in the real world is sometimes credited to US theoretical physicist Richard Feynman, who as long ago as 1959 gave a talk entitled 'There's Plenty of Room at the Bottom'. Nanotechnology has been developing steadily ever since, with thousands of patents filed and several Nobel Prizes awarded for work done in the field. Nanotechnology has a significant role to play at the coal face of numerous key areas of human progress. It is an exciting technology generally, but may have particular efficacy when it comes to addressing AMR, given a fundamentally different mode of action to our existing antimicrobial treatments.

The other reason for cautious optimism is the extent to which industry, governments, regulators and NGOs in many parts of the world have come together in recent years to work on solutions and to commit the capital required for finding them. I have already highlighted the existence of the AMR Action Fund. The fact that 24 biopharmaceutical companies have unilaterally come together and committed at least $1 billion in an attempt to address the problem is encouraging.

All over the world there are numerous initiatives focused on the problem and, seemingly, collaborating like never before. These include: ReAct, founded as long ago as 2005 with funding from the Swedish International Development Cooperation Agency (SIPA), Uppsala University and others; and the World Alliance Against Antibiotic Resistance, founded in Paris in 2011, and its sister publication *AMR Control*, first published in 2015.

I have already referenced the UK's *Independent Review on Antimicrobial Resistance*, commissioned by the British government in 2014 and published in full by Lord O'Neil and his team in 2016. The UK followed up in January 2019 with a new 20-year vision and five-year action plan on AMR, and the NHS and National Institute for Health and Care Excellence (NICE) launched an innovative antibiotic subscription model shortly thereafter in July 2019.

On the other side of the Atlantic, the US government and Food and Drug Administration has launched a series of initiatives and incentive schemes designed to address the problem. These have included the Generating Antibiotic Incentives Now (GAIN) Act, passed in 2012 and designed to promote the development of antibacterial and antifungal drugs to treat serious or life-threatening infections. The USA followed up on the 2012

GAIN Act with the 21st Century Cures Act in 2016, which helps innovative companies addressing smaller patient populations; the aptly named PASTEUR Act in 2020, specifically aimed at supporting the development of new antibiotics; and the DISARM Act, which aims to improve US Medicare reimbursement for antibiotics, and which is currently making its way through Congress.

Japan set up an AMR Task Force in 2015, and both it and China unveiled national AMR plans in 2016, the same year that the UN General Assembly held a 'High-Level Meeting on AMR', only the fourth time in the organization's history that a health topic has been discussed at the General Assembly.

All over the world, governments, NGOs, clinicians, biopharmaceutical companies, research scientists and investors are looking to tackle the problem head-on using the collaborative tools and processing power available nowadays.

More bad news – 'diseases of modernity' or 'modern plagues'

We have eradicated smallpox, massively reduced the impact of a raft of other previously horrific diseases such as cholera, hepatitis, malaria, polio, rabies, syphilis, tetanus, tuberculosis and numerous others, and enjoyed more than a century of reducing infant mortality, improving life expectancy and so much else besides. We take running water and sanitation for granted in the developed world, and we are working steadily towards the same position in the developing world.

But at the same time that we have achieved all of this, sadly and rather worryingly, a number of other 'diseases of modernity' or 'modern plagues' have been going in completely the opposite direction. These include: debilitating allergies such as hay fever, nut allergies and eczema, asthma, diabetes, epilepsy, inflammatory bowel disease (IBD) and irritable bowel syndrome (IBS), and a raft of other autoimmune diseases including rheumatoid arthritis, coeliac disease, myositis and lupus. Most of us are also more than familiar with the extent to which depression, other mental illnesses and obesity are on the rise all over the world.

We have seen the rate of asthma grow from about one child *per school* to a *quarter of all children* in less than a century. Similarly, peanut allergies trebled in the last decade of the twentieth century, and then doubled again in the five years after. Then there is diabetes, reasonably accurate records for which have been kept for a long time. Alanna Collen, in her book *10% Human: How Your Body's Microbes Hold the Key to Health and Happiness*, reports that – according to records from Massachusetts General Hospital in the US, spanning over seventy years until 1898 – there were only twenty-one cases of childhood diabetes out of nearly 500,000 patients. Formal record-keeping began just before World War II, allowing for the tracking of the prevalence of diabetes. In the US, UK, and Scandinavia, approximately 1 or 2 children out of every 5,000 were affected. By 1973, diabetes had become six or seven times more common than in the 1930s. In the 1980s, the upward trend flattened off to the current figure of about 1 in 250.

That is to say that levels of diabetes are up *between ten and twenty times* in less than a century. Coeliac disease is 30 or 40 times as common today as it was in the 1950s.

When it comes to rates of obesity, in the USA in the early 1960s, 13 per cent of the adult population were already obese, but by 1999 this proportion had more than doubled to 30 per cent. These numbers have continued to rise. Today it is estimated that more than 70 per cent of North American adults are overweight or obese. The same has been seen in many other developed-world countries including the UK and Australia.

Notwithstanding the fabulous progress we have made with so many of the infectious diseases which caused such misery to our forebears, all over the world, and particularly in the most developed parts of the world, a large number of other diseases and medical conditions are rising at an alarming rate.

Searching for causes

For some time, many researchers and clinicians assumed that these rises were primarily simply a function of improved diagnosis, possibly even *over*-diagnosis. If we test for things today which we didn't test for a few decades ago, or if our healthcare systems are better at diagnosing certain conditions or more inclined to go looking for them than was the case in the past, we will see significant 'growth' in those conditions.

By now it seems reasonably clear, however, that growth rates in such things are certainly not merely a function of improved rates of diagnosis. When researchers adjust for that factor, these many diseases of modernity really are significantly on the rise.

Not long after we triumphed over infectious disease, we seem to have started losing the battle with chronic illness. It is becoming increasingly clear that these two things are likely to be linked. This seems to be a case of squeezing the proverbial balloon. The more we have squeezed at one end with antimicrobials and vaccines to combat infectious disease, the more the balloon has bulged at the other end with an explosion in a raft of other chronic health problems.

A key clue to a causal relationship here lies in the differing experiences of the growth trajectory of such chronic conditions in the *developed* world as against in the *developing* world. These diseases of modernity are very significantly less prevalent in the developing world than in the developed world. As one study published by Okada and colleagues in the journal *Clinical and & Experimental Immunology* has put it: 'The geographical distribution of allergic and autoimmune diseases is a mirror image of the geographical distribution of various infectious diseases.' Collen puts it succinctly: 'A great accumulation of evidence points to the correlation between chronic diseases and affluence.'

It seems clear that this is a function of affluence rather than genetics, too. As the Okada study cited above puts it:

> the incidence of diabetes is sixfold higher in Finland compared to the adjacent Karelian republic of Russia, although the genetic background is the same. Additionally, migration studies have shown that offspring of immigrants coming from a country with a low incidence acquire the same incidence as the host country, as rapidly as the first generation. This is well illustrated by the increasing frequency of diabetes in families of immigrants from Pakistan to the United Kingdom or the increasing risk of MS in Asian immigrants moving to the United States. The prevalence of systemic lupus [...] is also much higher in African Americans compared to West Africans.

Even *within* a given country, numerous studies have shown a correlation between wealth and chronic illness. After (poor) East Germany reunified with (wealthy) West Germany in the early 1990s a study showed that the richer West German children were twice as likely to have allergies than their East German compatriots.

Put simply, the more successful a nation state has been at eradicating infectious disease, and the wealthier a given population is, the more of a problem they have with chronic illness. It seems increasingly clear that these are two sides of the same proverbial coin.

It's all about those microbes

The Human Genome Project has estimated that humans have between 20,000 and 25,000 genes. This is a tiny fraction of the number of genes that are found in our *microbiome*, that is to say in the trillions of bacteria that lurk in us and on us. As reported in *ScienceDaily* in 2019, a team of scientists at Harvard Medical School 'set out to estimate the size of the universe of microbial genes in the human body, gathering all publicly available DNA sequencing data on human oral and gut microbiomes. In total, they analysed the DNA of some 3,500 human microbiome samples, of which more than 1,400 were obtained from people's mouths and 2,100 from people's guts.' They found that there were nearly 46 million bacterial genes from those 3,500 samples. Extrapolating from that data, they suggested that, in total: 'There may be more genes in the collective human microbiome than stars in the observable universe.'

In recent years it has become increasingly clear that our microbiome and virome have an extremely important role to play in our health overall. As that Harvard study put it:

> Mounting evidence has revealed the role of [. . .] microbes as powerful modulators of disease and health [...] linked to development of conditions ranging from [...] dental caries and gut infections to [...] chronic inflammatory bowel disease, diabetes and multiple sclerosis.

Falling rates of infectious disease have correlated with rising rates of chronic illness all over the world. It now seems clear that the reason for this has to do with the extent to which our infectious disease fighting technologies and our lifestyles have *affected the balance of our microbes*.

There are clues that this is the case everywhere we look. Arguably of most significance is the fact that some 80 per cent of our immune system is found in the gut. Crucially, our gut is also where an estimated 95 per cent of our microbiota can be located (our own personal bacteria). There is a causal relationship between the quantity and quality of our gut 'flora' and the efficient functioning of our immune system and, by extension, our health overall.

Dysbiosis

'Dysbiosis' is the term used to describe a compromised or imbalanced microbial community in the human gut. Research conducted over the last several years has implicated dysbiosis in disease after disease, including essentially all of these chronic diseases of modernity as well as cancer, obesity and a wide range of mental health conditions, including depression.

The relationship between our gut health and our health overall, particularly our mental health, is sometimes described by clinicians, scientists and nutritionists as 'Gut and Psychology Syndrome', or GAPS. GAPS is a theory that suggests a link between the health of the digestive system and the development of neurological and psychological disorders and a number of other conditions, including depression, anxiety and schizophrenia. When we consider the list of things which cause dysbiosis, we find the causal link between reducing rates of infectious disease and rising chronic illness and can see clearly why those illnesses are so often diseases of affluence and modernity.

Specifically, risk factors for dysbiosis include: a poor diet, including too much sugar (and alcohol); the 'accidental' consumption of chemicals such as those commonly found in pesticides and elsewhere in the consumer products of our modern, industrialized world; excessive use of antibiotics, and high levels of stress. As should be abundantly clear from this list, all of these factors correlate with rising development and affluence, particularly the use of antibiotics, for example.

Another key related factor is sometimes described as 'the hygiene hypothesis'. As leading US healthcare group the Mayo Clinic explains: 'The hygiene hypothesis proposes that childhood exposure to germs and certain infections helps the immune system develop. This teaches the body to differentiate harmless substances from the harmful substances that trigger asthma. In theory, exposure to certain germs teaches the immune system not to overreact.' This is thought to be why 'American children living in poverty are historically less likely to suffer from food allergies and asthmas than their wealthier counterparts.'

It is also why numerous studies have found rates of allergies and other chronic conditions are lower for children growing up in rural settings and on farms than for their urban-dwelling peers. As a 2015 *Science* article by Martijn J. Schuijs and colleagues has put it: 'People who grow up on dairy farms only rarely develop asthma or allergies. This is probably because, as children, they breathe air containing bacterial components, which reduce the overall reactivity of the immune system.' This is also very likely to be the reason that children in many migrant populations suffer increased rates of all of these sorts of health conditions as we have already seen.

Another increasingly well-documented risk factor for all of the above is the rise of children being delivered by Caesarean section. Globally, rates of Caesarean section have doubled in the last decade and a half to 21 per cent, leading one of the world's oldest peer-reviewed medical journals, *The Lancet*, to describe what they deem to be the 'overuse' of C-section as nothing less than an 'epidemic'. One of the reasons for their position is because, as a 2020 study by Gyungcheon Kim and colleagues published in *Frontiers in Microbiology* explains, Caesarean delivery is associated with 'the perturbation and delayed maturation of gut microbiota in early life, which in turn has been associated with increased risk of childhood obesity, asthma, immune diseases and infectious outcomes'.

It is becoming abundantly clear that a modern, affluent, excessively sterile and overly medicated lifestyle, particularly when it involves the use of large quantities of antibiotics, can significantly compromise our gut flora and, by extension our immune system and our health. This is highly likely to be one of the key common causal factors leading to the rise in so many chronic health conditions all over the world and in the developed world in particular.

That is the bad news. The good news, however, is that we seem to have finally worked out what the problem is and, increasingly, what we might be able to do about it.

Of course, we don't want to lose so many of the benefits of modernity and, most particularly, our ability to fight infectious disease with antimicrobials and vaccines. As we have seen, giving up such things would take us back to a world where hundreds of millions of us might die unnecessarily. Surgery would become prohibitively dangerous, for example. Equally, however, we don't want our societies to continue to endure this inexorable rise in diabetes, obesity, depression and so much else besides, which have come as a natural consequence of that modernity and of our use of vaccines and antimicrobials.

Happily, it is highly likely that we will be able to 'have our cake and eat it'. Part of the reason I believe this to be the case has to do with the fact that the wealthiest within society have shown themselves most able to avoid much of the chronic disease epidemic. This is because such individuals so often have access to the latest and best health information, the best healthcare, and the freedom in terms of time and money to make the healthiest choices around diet, exercise and supplementation for example. Economic growth and technological development will make these advantages available to an increasing percentage of the population, just has been the case with every other element of human progress since the Industrial Revolution, whether the radio, car, plane, television, personal computer or smartphone.

If there is one primary solution to so many of these diseases of modernity, it seems likely to lie in our ability to deal with dysbiosis, particularly in our children as their gut flora develops. We need to ensure that our microbiota and our immune systems and those of our children, do not become excessively compromised by these various risk factors of modernity. Encouragingly, to do that we may need nothing more than better information and a willingness to take consistent action and implement habits which reduce the risk of dysbiosis and the many health problems it can inflict.

For many, repairing a compromised gut microbiome with all that this implies for the effective functioning of their immune system and a reduction in chronic disease, may 'only' require better diet and nutrition, supplementation with probiotics and a number of lifestyle choices around things like sugar consumption, sleep, exercise and the ability to deal with stress. At present, too few people have sufficiently good information about such things, but I would argue that this is changing and will likely do so at an exponential rate going forwards, in large part thanks to the biotech and tech industries.

In summary

The last century or so has been something of a roller coaster in terms of health. While vaccines and antimicrobials have been able to reduce, and in some cases eradicate, many of the age-old diseases that have curtailed and blighted human lives, affluent modernity seems to have stimulated previously less prevalent conditions, creating new epidemics from diabetes to depression.

In the next chapter we will look at how technological progress is poised to deliver many of the substantive changes we need to make in order to stop and then reverse the rise in so many of our modern health problems and diseases of modernity and the role played by the biotech industry in particular.

6
Biotech and your health

So far in this part of the book we've looked at the extent to which modern medicine has triumphed over infectious disease, primarily as a function of antimicrobials and vaccines. We then examined the causal link between that technological progress and an explosion in a wide range of diseases of modernity and affluence that have resulted. In this chapter, we'll look at the many ways in which we might address those diseases and health problems and the extent to which our ability to succeed here will be a function of the rapid development of a number of related (bio)technologies.

All about the apps

In recent years, there has been an explosion of interest in 'wellness', diet, nutrition and exercise. After the inevitable hiatus caused by COVID and lockdowns, in the USA the fitness industry is forecast to more than quadruple between 2020 and 2028, from just over $100 billion per annum of revenues to some way more than $400 billion.

The fastest-growing segment of that industry is in 'online and digital' which is expected to grow at more than 30 per cent per annum in the next few years. Today anyone who owns a smartphone has access to better, cheaper and more conveniently available information about health, fitness, diet and nutrition than at any other time in history through a raft of quality 'apps'. There is rapid growth in the adoption of such apps and early signs that they can increase adherence to a given exercise programme, diet or other habit which can have a positive influence on the user. One study by Manel Valcarce-Torrente et al. published in October 2021 'concluded that 72 per cent of users considered that the use of exercise apps motivated them towards the achievement of their daily physical practice goals'.

The same phenomenon is at work with the explosion of interest in meditation and mindfulness apps. Numerous studies have suggested the likelihood of significant health and mental health benefits from meditation. These include, reducing stress, anxiety and depression, managing pain and addiction, improving sleep and decreasing blood pressure. Other studies have suggested that mindfulness has a role to play in the treatment of numerous diseases, even ones as serious as cancer, where reduced stress and improved sleep alone, for example, are clearly of benefit to a patient.

In their 2017 book *The Telomere Effect* Drs Elizabeth Blackburn and Elissa Epel explain the role played by 'telomeres' in health and most particularly in the rate at which we age. As the National Human Genome Research Institute explain: 'A telomere is a region of repetitive DNA sequences at the end of a chromosome. Telomeres protect the ends of chromosomes from becoming frayed or tangled. Each time a cell divides, the telomeres become slightly shorter. Eventually, they become so short that the cell can no longer divide successfully, and the cell dies.' In their book Blackburn and Epel show how telomeres 'have a direct effect on how quickly or slowly we age' and point to a 'growing body of evidence [that] suggests that meditation training may have a range of salubrious effects, including improved telomere regulation'. It is also interesting to note that there is more than likely a link between 'inflammation, telomere length, gut microbiota and psychiatric disorders', as proposed by a study published in the *British Medical Journal* in January 2020.

There is more evidence for the merits of meditation overall. Tim Ferriss is the *New York Times* best-selling author of several excellent books and host of one of the top business podcasts in the world, *The Tim Ferriss Show*. Over the last several years he has conducted more than 700 interviews with an extraordinary cross-section of many of the world's topmost performers, including politicians, writers, philosophers, billionaire entrepreneurs and businesspeople, Hollywood actors, musicians, research scientists, clinicians, athletes and investors. His back catalogue of interviewees includes people such as Richard Branson, Arnold Schwarzenegger, Eric Schmidt, Mark Zuckerberg, Kevin Costner, Hugh Jackman, Madelaine Albright and Ray Dalio, to name just a few.

His 2016 book *Tools of Titans* includes contributions from several dozen of his most famous and celebrated guests and features a section on habits which are common to many of those individuals. The first habit highlighted in that section is the fact that 'more than 80 per cent of the interviewees have some form of daily mindfulness or meditation practice'.

I confess I found this statement fascinating given the astonishing range of activities represented by this group of people. Given increasing evidence for the merits of meditation overall, it is positive that the proportion of adults practising some form of meditation or mindfulness has more than tripled since 2012. *Calm*, the leading smartphone app which teaches people how to meditate, has been downloaded by more than 135 million people and was Apple's 'App of the Year' in 2017. Its main rival, *Headspace*, is not far behind, having been downloaded 70 million times.

All over the world, people are increasingly focused on their health, fitness, diet, nutrition and mental health. Hundreds of millions of us are already helped in this endeavour with access to a wide range of inexpensive and even free smartphone apps and other online services which, as we have seen, can provide the information we need and increase adherence to and compliance with a given programme or habit.

It might seem naive to suggest that we might address problems as serious as diabetes, obesity or depression with a rise in the number of people using health and fitness apps and improving their diets. This position, however, underestimates the power of exponentials over time and the power of better science, better information and better tools to effect meaningful change. Science is increasingly giving us the answers and the tools we need to make that change and giving it to an increasing percentage of the global population, too.

There is mounting evidence that smartphone apps can be a powerful catalyst for improving people's health, fitness, diet and other key habits. Another 'convergence device' that is probably not too many years away from being widely adopted and which may be even more powerful in that respect is that of 'augmented reality', or 'AR', glasses (spectacles). Many of the world's leading technology companies see these as the natural progression from smartphones. Alphabet (Google), Apple, Meta (Facebook), Microsoft, Snapchat and others are investing billions of dollars in their development.

Advanced versions of such products are already being used in surgery, design and architecture. In the relatively near future, however, they may become as widely adopted as smartphones and possibly even replace them. This is certainly the view held by people leading the research and development divisions of several of the world's largest tech companies. Balaji Srinivasan is a partner at top US venture capital firm Andreessen Horowitz and was previously named one of *MIT Technology Review*'s 'Innovators Under 35'. He uses the phrase 'optimalism' to describe the potential in such technologies.

Not too many years from now, voice-activated AR glasses, combined with a wearable technology such as a wrist band or ring, may be able to provide us with a real-time heads-up display which can monitor every aspect of our health and provide us with an extremely convenient and regular series of calls to action in order to make the right choices around diet and exercise and even help us reduce our stress levels in a myriad of ways.

The wide adoption of these sorts of technologies and steady improvement in our access to information and our implementation of best practice can and will go a long way to slowing and then even significantly reducing the instance of these many diseases of modernity, just as our vaccines and antimicrobials did with infectious disease.

Exponential science

Crucially, however, our ability to deal with such things won't just be about lifestyle improvements engendered by better tech. It is clear that our microbiome, virome and genome are absolutely instrumental in our health. We'll look at each of these in more detail later on, but it's worth explaining these terms here. As a reminder, our *microbiome*, or simply our 'biome', is the collection of microbiota (microorganisms) living on and inside our bodies, such as bacteria and fungi, together with their genes. We also have a *virome* (a part of the biome) – which is all the viral cells we carry. These account for another 10 trillion cells or so within each of us. In addition, we all then have our own personal *genome*, too – our genetic make-up. Crucially, our biome, virome and genome are highly differentiated and incredibly personal to each of us.

Our health or ill-health is a result of the extraordinarily complex and individual interplay of those millions of bacterial genes, trillions of bacterial and viral cells and thousands of human genes. It is also a result of our environment, diet and lifestyle choices and stress levels. Here, too, there is cause for optimism simply because of the exponential progress in our ability to analyse and understand such complexity. Eric Schmidt was the CEO of Google from 2001 to 2011. As he has put it:

> The chemist wakes up in the morning and says, 'Let's try the following seven compounds.' They try the seven compounds, none of them work. And at five o'clock, they go home to have dinner and think, watch television, and the next morning they think of another seven. Well, the computer can do a hundred million in a day. That's a huge accelerant in what they're doing.

This matters. The biotech industry today has innovative analytical tools, processing power and the ability to collaborate like never before, and such things are also improving at an exponential rate. With each passing month our ability to understand the enormous complexity in our genome, biome and virome increases as does the likelihood that we can develop effective solutions for our most intractable health problems as a result. This is further compounded by the steady development of a wide array of apps and other technologies such as wearable devices which will very likely help an increasing number of us to make the right choices for our health.

A personal tale – a compromised biome

We have already seen just how important the role played by our biome is for our health and how fundamentally damaging *dysbiosis* – a compromised microbiome – can be. I have a very personal interest in the role played by our microbiome in particular because in 2008 I was diagnosed with a condition called ulcerative colitis (UC). For those who may not have heard of UC (I hadn't at the time I was diagnosed), along with Crohn's disease, it is one of the two inflammatory bowel diseases (IBDs).

In 2015 the US Centers for Disease Control and Prevention estimated that 1.3 per cent of adults in the USA (fully three million people) suffer from IBD, and showed that the instance of it has been increasing for all the reasons given in this part of the book. UC and Crohn's disease can both be serious and debilitating illnesses. The symptoms can range from pretty mild (although potentially entirely embarrassing) to sufficiently serious to be life threatening and to necessitate serious medication and even life-altering, disfiguring surgery.

From 2008 to about 2018 my condition was debilitating and unpleasant and the cause of a great deal of stress. The reason I mention it is because my journey back to being essentially symptom free, which I am fortunate to enjoy today, involved my learning a great deal and most particularly about the importance of the microbiome in health. This was helped in no small part by the fact that I happened to be working in the life sciences industry and lucky enough to meet companies working in the area. I have also been able to make use of many of the technologies and treatments I have referenced over the previous few pages.

The silver lining on the awful black cloud of having had such a serious chronic condition has been that I have learned a fair bit about the microbiome, about how important the gut is for health overall and, as a corollary to that, some really key things about diet, supplementation and certain other lifestyle choices which I believe are, sadly, still poorly understood or, at the very least, poorly implemented by many people. My reading also suggests to me that many of the things I have learned as a result of suffering from UC are broadly applicable to a wide range of other chronic health conditions and many of these diseases of modernity.

Your genome

Most people will know that our own personal genetic make-up will determine our skin, hair and eye colour and lots else besides when it comes to our physical appearance. It also has a significant role to play in illness, as you might expect, and in whether we are naturally predisposed to suffer from a given disease or health problem.

As an example, readers may remember that the Hollywood film star Angelina Jolie chose to have elective surgery some years ago to reduce her risk of breast and ovarian cancer. She carried something called the BRCA1 gene (BReast CAncer gene 1), which scientists have established significantly increases the risk of both of those cancers – hence her decision to have the surgery as a 'pre-emptive strike' against that reality. This is a pretty high-profile example of the fact that your own genetic make-up is instrumental in a great deal to do with your health.

It also seems reasonably clear that your genome has a role to play when it comes to mental health. The University of Oxford's Department of Psychiatry cited a study published in May 2020 revealing 'a common pattern of connections in the brains of people whose genes predispose them to mental health problems. Findings show that brains "wired up" in this way are associated with not just one but a whole range of mental health conditions, including schizophrenia [...] depression, anxiety, and bipolar disorder.' This relationship between our genes and mental illness is now the subject of a great deal more research using the increasingly powerful and sophisticated analytical tools now available to scientists and clinicians.

Crucially, although we may be genetically predisposed to be more at risk from certain health problems, diseases or mental health disorders, there is seemingly quite a lot we can do about those risks. We can change our diets and, in particular, work to improve our microbiome.

'Nutritional genomics' – and beyond

It is becoming increasingly clear that our genome has an important role to play in diet and, in reverse, our diet may affect our genome, too. This is the focus of the emerging field of 'nutritional genomics'. The field is split into 'nutrigenomics' and 'nutrigenetics'. Joanna Janus of the PHG Foundation at the University of Cambridge explains: 'Nutrigenomics assesses how nutrition affects genome regulation. Nutrigenetics investigates specific genetic variants that regulate nutritional processes.' Put simply – our genetic make-up can impact our ability to assimilate nutrition, and the quality of our nutrition, particularly over long periods of time, can impact the functioning of our genes. There is a circular relationship.

The idea that our genes might affect our relationship with food has been around for several decades. It was initially not taken particularly seriously by the medical or scientific establishments. There were many slightly fringe, self-proclaimed nutrition 'experts' touting one diet or other based on this broad worldview. Relatively well-known examples include diets based on your blood type which were particularly popular in the 1990s with proponents selling many millions of books on the subject. The basic thesis was that certain blood type groups respond better to a vegetarian diet and others to a diet heavier in meat and fish. Over thousands of years of evolution, proponents of this thesis suggested, human beings in different parts of the world adapted to their surroundings and 'optimized' *genetically* to the foods that were most readily available to them in their immediate vicinity. One of those genetic adaptations, they argued, was their blood type.

Northern Europeans and hunter gatherers tended to have a high protein diet given how much fish and meat they consumed as a natural function of the way they lived and of their environment. The implication was that such people would then be 'healthier' consuming that diet, given they had adapted to it over thousands of years. People in primarily agrarian societies who secured the majority of their calories from plants and grains, in places like the Indian subcontinent for example, would be better served by a vegetarian or vegan diet.

Many scientists and clinicians have subsequently found fault with the blood-type thesis specifically, based on the analysis of large datasets, and suggested that there is little or no scientific evidence to support eating based on your blood type. However, even if the blood-type diet may have been on the wrong track, by now there seems little question that the complex interplay between your genetic make-up and your diet will certainly be consequential for your health and for working out which diet is 'right' for you.

Other memes which have emerged around a genetic basis for diet include ideas such as the supposed remarkable health of certain ethnic groups, particularly those still 'living as they always have' – a dangerous and simplistic idea.

Two particular examples of this which are often cited and seem to have entered our collective consciousness over time are the Maasai in East Africa and the Inuit peoples of northern Canada, Greenland and Alaska.

Countless articles and more than a few compelling-sounding television documentaries have highlighted the fact that both of these two peoples consume diets which are very high in fat and protein and relatively low in vegetables or fruits, yet they exhibit much lower rates of the illnesses those of us in the developing world invariably assume would result from such a diet. Such findings are often used by proponents of high protein, high fat diets such as the Atkins or 'carnivore' diets to support their position.

At the other end of the argument, advocates of vegetarian or vegan diets have their own examples of remarkably healthy peoples to point to. Well-known examples include the Okinawans in southern Japan and the rural Chinese. As a 2009 study of the Okinawans by Craig Willcox and colleagues has put it, they 'are known for their long average life expectancy, high numbers of centenarians, and accompanying low risk of age-associated diseases'. That same study highlights the fact that the Okinawan diet is 'vegetable and fruit heavy (therefore phytonutrient and antioxidant rich) but reduced in meat'.

Another famous piece of research was conducted in the 1980s by academics from Cornell and Oxford universities working with the Chinese Academy of Preventive medicine. According to the Center for Nutrition Studies, that study:

> embarked upon one of the most comprehensive nutritional studies ever undertaken known as the China Project. China at that time presented researchers with a unique opportunity. The Chinese population tended to live in the same area all their lives and to consume the same diets unique to each region. Their diets (low in fat and high in dietary fiber and plant material) also were in stark contrast to the rich diets of the Western countries. The truly plant-based nature of the rural Chinese diet gave researchers a chance to compare plant-based diets with animal-based diets.

The study came down heavily in favour of plant-based nutrition. 'The evidence' from the Maasai and the Inuit and 'the evidence' from the Okinawans or rural Chinese go some way to explaining why it is possible to find numerous books which 'scientifically prove' that if you want to maximize your chances of being healthy you should almost certainly be vegetarian or vegan and plenty of others which 'scientifically prove' that you should do the precise opposite, fill yourself full of meat and fat and limit your intake of carbohydrates. How can there be such confusing and conflicting advice?

In recent years, the science suggests that the answer is actually relatively simple: there is a key missing link – a factor that hasn't been given sufficient consideration. Crucially, few of the analyses of these diets have taken account of the role played by bacteria and viruses – of the biome, virome and, to a certain extent, of the genome too when it comes to our diets, either sufficiently or at all.

Scientists have become increasingly interested in the role played by our genes when it comes to diet, as we have seen. As a March 2021 article published by Healthline puts it: 'If there's one thing the last several decades of nutrition research have proven, it's that *there's no one-size-fits-all diet.* While many factors are at play, one reason certain eating plans work for one person but not another may have to do with our genetics.'

But there is a great deal of evidence by now that it goes much further than genetics alone. Our environment, our genes, our diet and our lifestyle will impact those trillions of bacterial and viral cells in each of us and this, in turn, will impact our health. As reported in *ScienceDaily*, a 2019 University of Pennsylvania study found: 'Our microbiome [...] reflects the way we live. If we own a pet, we likely share microbes with them. If we eat meat, the microbiome in our intestines may look different from that of a vegan.' If we grow up on a farm, our microbiome will be fundamentally different from that of someone who grows up in the centre of Manhattan. If we grow up in southern Japan, it will be very different to what it might be had we grown up inside the Arctic Circle.

Crucially, our biome then has a fundamental role to play in how we assimilate nutrients – in how we react to food. The fact that people in Okinawa, Alaska and Tanzania can consume such wildly different diets yet still have lower rates of all sorts of Western diseases is a function of this reality. These peoples have different health outcomes on different diets partly because they are different genetically, but arguably more important than that, because they have different microbes – inside them and also around them in their environments – and the same is very likely true of all of us. Our diet and our genes are important, but arguably even more so is the complex interplay of such things with the composition of our microbiome.

The obesity example

As an example of just how compelling these ideas are, it is perhaps instructive to look at obesity.

Most of us will be familiar with the debate that has raged endlessly for decades between 'fattists' – who maintain that anyone with a weight problem simply lacks will power, eats too much and exercises too little – and people at the other end of the argument – often those suffering from obesity themselves, or close to someone who is – who insist they have a medical problem, poor metabolism or some other explanation for why they're incapable of controlling their weight.

In August 2019 veteran British BBC news presenter Michael Buerk famously caused a storm of controversy in the UK when he wrote in one of the largest-circulation British weekly magazines: 'You're fat because you eat too much.' The emotional reaction to his comments illustrated just how deep this debate goes in our society. Michael Buerk and others like him believe that obesity is nothing more than what results if you eat too much and exercise too little. Millions of overweight people disagree and maintain that they're powerless to do anything about their condition.

Increasingly, the science is showing us that both camps are probably wrong. First, obesity is *not* a simple function of calories in and calories out. Secondly, obese people *do* have the power to do something about their condition – but not if they focus on diet and calorie restriction alone, as has been the convention for most of our lifetimes and the 'solution' attempted by the vast majority of people who are overweight.

Whether you become obese or not has an enormous amount to do with the role played by *bacteria* and *viruses* – by your microbiome – and is *not* just a function of what you eat and how much you exercise. As a July 2021 study by Bing-Nan Liu and colleagues published in the *World Journal of Gastroenterology* explained: 'Obesity is closely related to the gut microbiota. The study of the gut microbiome provides a basis for the reconstruction of the gut microbiota of obese patients.' The study discussed the characteristics of the gut microbiota in obesity, *how* the gut microbiota induce obesity, and the relationships between environment and genetics and the gut microbiota in obesity, and then went on to conclude:

> Dysbiosis of the gut microbiota has been shown to be closely linked to obesity. Many gut microorganisms have been identified to be related to obesity. They induce the occurrence and development of obesity by increasing host energy absorption, increasing central appetite, enhancing fat storage, contributing to chronic inflammation, and regulating circadian rhythms.

This is extremely important. Alanna Collen begins the second chapter of her 2015 book *10% Human* by sharing a fascinating little story about a bird called the common garden warbler. Along with many other species of bird, warblers migrate as far as 4,000 miles from Europe to sub-Saharan Africa in the northern hemisphere's winter. According to Collen, before embarking on their remarkable journey, these small birds prepare themselves for the demanding flight and the scarcity of food along the way by accumulating fat. Within just a few weeks, the warblers double their weight, increasing from a slender 17 grams to a notably hefty 37 grams. In human terms, they become extremely overweight. During each day of the pre-migration feeding frenzy, a garden warbler gains approximately 10 percent of its initial body weight – akin to a 10-stone man adding a stone to his weight every day until he reaches 22 stone.

Crucially, when researchers analysed how many extra calories the birds consumed they 'realised that the additional food the birds were consuming did not fully account for the weight they were managing to gain'. The really extraordinary point Collen makes, however, is what happens to *captive* garden warblers. In the lead-up to migration at the end of summer, these confined birds continue to gain weight, becoming excessively obese in readiness for a journey they will never undertake. Interestingly, precisely when the wild warblers reach their destination, the captive warblers lose their surplus fat entirely, despite never embarking on the 4,000-mile flight.

To me this is an astonishing example from the animal kingdom that suggests there may be a great deal more to weight gain (and loss) than calories consumed.

The story unfolds still further as Collen cites work done by Fredrik Bäckhed and Jeffrey Gordon and their colleagues of Gothenburg University, Sweden, and Washington University in St Louis, Missouri, respectively. As long ago as 2004 these scientists investigated whether changing microbes in mice might have an impact on their weight. The result of one of their experiments was that certain of the mice being examined saw 'a 60 per cent increase in body weight in fourteen days. And they were eating less.' Mice seeded with gut flora that predisposed them to obesity saw enormous changes in their weight as against the ones which had a 'normal' or 'thin' microbiome who saw no such weight increase, even when all of the mice were consuming the same amount of food – or, in some cases, where the obese mice were even consuming less food. One of the team, PhD student Peter Turnbaugh, 'calculated that the mice with the obese microbiota were collecting 2 per cent more calories from their food'.

Although this may not sound like much, Collen goes on to highlight what a difference that sort of change could make in a human being. As she says:

> Let's take a woman [...] who weighs 62 kg. [...] She consumes 2,000 calories per day, but with an 'obese' microbiota, her extra 2 per cent calorie extraction adds 40 more calories a day. Without expending extra energy, those further 40 calories per day should translate, in theory at least, to a 1.9 kg weight gain over a year. In ten years, that's 19 kg. [...] All because of just 2 per cent extra calories extracted from her food by her gut bacteria.

This is a 'smoking gun' if ever there was one in the debate. A person whose gut flora 'helps' them extract only two per cent more from the calories they consume each day could end up being morbidly obese as compared to someone with a different microbiome who experiences no such impact from their gut flora.

We have already seen from the study published in the *World Journal of Gastroenterology* that someone's microbiome can increase their appetite levels, too. It seems quite clear by now that your own very personal set of microbial 'guests' will have a significant role to play in how easy or hard you find it to avoid becoming fat.

Interestingly, it seems that viruses may have a part to play here, too. One virus in particular, Adenovirus 36, has already been linked to obesity. Maps charting the spread of the 'obesity epidemic' in the USA over the past three or four decades give much the same impression as an infectious disease sweeping through the population. Of course, if bacteria and viruses have a key role to play here, then this would make sense intuitively given both can be passed on from one person to another. A married couple will tend to have similar lifestyle habits in terms of what they eat and the extent to which they exercise, but studies looking at this phenomenon controlled for that factor. The causal relationship was a function of something more than just similar lifestyles.

It is also worth noting that there seems to be a fair amount of heritability when it comes to your microbiome. This is yet another factor which goes some way towards explaining why obesity so often runs in families. The children of obese adults may be genetically predisposed towards obesity to begin with. This may then be compounded further by the risk they have inherited a compromised microbiome from their mother, and then further still should they inherit their parents' habits and lifestyle choices, of course. More often than not, a child will 'inherit' their parents' environment, too, assuming that the family don't move away to an entirely new area shortly after a child is born, which means they inherit the microbes that are found in that environment too of course.

While all of the above will increase the risk that an individual becomes obese, and while this problem will compound over time as the instance of obesity in our society overall increases, this is not to say that such individuals are powerless to reverse matters, particularly if they and their medical practitioners or nutritionists begin to understand better the crucial role played by their microbiome and by dysbiosis. Notwithstanding what a complicated, multifactorial condition obesity is, there is an emerging body of evidence that suggests that one of the key solutions to the problem might come from placing our focus at least as much on the role played by the microbiome in future as we have done on the role played by diet and exercise in the past. In fact, it should be front and centre in our arsenal of solutions to combat the obesity epidemic and, indeed, a large number of other growing health problems. There is mounting evidence that a compromised microbiome, or 'dysbiosis', is implicated in a wide range of diseases, including obesity. Logically, therefore, it would seem that addressing dysbiosis should be considered as a pretty key component of any treatment plan for obesity and a great deal else besides.

A compromised microbiome can increase the calorific impact of the food we eat, increase our appetite structurally, and decrease our ability to assimilate key nutrients, which will also make us feel hungry as a result. This reality goes some way to explaining why dieting alone so often fails people suffering from obesity.

It doesn't matter how 'healthy' your diet is if you have poor-quality gut flora and your ability to derive benefit from that diet is significantly challenged as a result. Similarly, just like those garden warblers and laboratory mice, if your microbial passengers are calibrated such that the calorific output from what you eat is structurally higher than what it would be for a person of normal weight and with a healthier microbiome, it is entirely possible that you will continue to put on weight even if you were to eat fewer calories than they do.

At first glance, this might seem incredibly unfair to anyone struggling with their weight and really rather depressing. The good news, however, is that if dysbiosis is as instrumental as it increasingly appears to be, this actually gives an obese individual far more agency over the problem than the approach we have tended to employ to date – that is, focusing on diet and calorific restriction alone.

Repairing the microbiome

Crucially, not only are we increasingly aware of the role played by dysbiosis here, but we are also increasingly aware of what to do about it. If a compromised microbiome is one of the key underlying causes of obesity, the good news is that nowadays we are beginning to develop a pretty good idea of how to repair it. In the following section, we will look at this idea in more detail.

The 'ancient wisdom' of fermented foods

In July 2021 the Stanford School of Medicine in California published the results of a small clinical trial. They found that 'a 10-week diet high in fermented foods boosts microbiome diversity, improves immune responses [...] and decreases molecular signs of inflammation.' As Dr Justin Sonnenburg, associate professor of microbiology and immunology put it 'This is a stunning finding. [...] It provides one of the first examples of how a simple change in diet can reproducibly remodel the microbiota across a cohort of healthy adults.'

Health food fans have long stressed the supposed merits of a wide range of fermented foods, whether kefir from Turkey, sauerkraut from Germany, natto beans and miso from Japan, kimchi from Korea, tempeh from Indonesia, kombucha, originally from China, the fermented whale and shark meat consumed by peoples in the Arctic Circle, and even the notoriously disgusting fermented Baltic herring known to the Norwegians as *rakfisk* and to the Swedes as *surströmming*.

As can be seen, such foods have developed in different places and times all over the world and across many cultures (as have alcoholic drinks such as beer and wine both of which can increase microbial diversity in the gut when consumed in moderation). Although a great deal more research needs to be conducted, it seems more or less clear by now that such foods are likely to have a positive effect on our gut flora which, by extension, will have a positive effect on our health. It is also clear that consumption of such foods has tended to reduce over time to be replaced by a more 'modern' diet of ready meals and processed foods – until a resurgence, in recent years, among affluent health-conscious Western consumers at least.

It seems clear then that fermented foods of all kinds are probably a good thing when it comes to addressing obesity and, indeed, a raft of other diseases and most particularly those diseases of modernity whose inexorable rise in the last century seems highly likely to be related to the deterioration of our collective microbiota. That said, there is clearly significant variability in the efficacy of such a wide range of different foods. It is also hard to establish any kind of accurate 'therapeutic' dose with so many of them to choose from and with relatively few large academic studies conducted to date. Should a sick or obese person suffering from dysbiosis have a litre of kefir or kombucha a day, 100 grams of natto beans, or one serving of kimchi or ensure they consume some mix of all of the above in any given week?

Any such approach will necessarily be rather hit or miss, particularly when you consider that this implies adding yet more complexity to an obese or sick person's diet. It is already well established that 'compliance' – actually sticking to a diet – is one of the great challenges for clinicians dealing with obesity and, indeed, so many other health conditions where a nutritional plan might be brought to bear. Adding yet another handful of things to remember to consume each week simply increases the probability that a given patient fails to stick to the plan.

A simpler approach, therefore, may be to ensure that a given patient takes a sufficiently powerful 'probiotic' which can help repair their gut flora. A probiotic is simply a product which contains live microorganisms which it is hoped might provide health benefits, most particularly in dealing with dysbiosis. Such products are most often described as containing 'friendly bacteria' and are widely available, most often in liquid, powdered or tablet/capsule form.

Symprove

Some pages ago I mentioned my ulcerative colitis diagnosis in 2008. As I began to understand the likely importance of dysbiosis in my condition, I tried a number of probiotics and probiotic foods such as kefir for several years with little or no result. My symptoms persisted.

Then, quite by chance, sometime in 2014 or 2015, I stumbled upon a liquid probiotic product called Symprove. Over several weeks of using Symprove my symptoms relented and after probably about two years of using it on and off, and of making some other key lifestyle choices, I became entirely symptom free. It has now been several years since I have experienced any of the symptoms ordinarily associated with IBD (colitis or Crohn's disease). I appreciate that my own personal experience is anecdotal and entirely unscientific. I am only one person after all – an 'n of 1' in the vernacular of a clinical trial. But by now I don't think this matters.

Because of that personal experience and my belief in the product, it struck me that I should look into the company a bit more. I was particularly motivated to do this because it seemed that almost no one had heard of it. Not long after my colitis diagnosis I joined the Crohn's & Colitis UK group on Facebook. At the time of writing this group currently has nearly 58,000 members. When I first discovered Symprove I found that almost no one in that group had heard of it. There were thousands of people in the group, many of whom were suffering terribly from such an awful condition, taking powerful drugs with serious side-effects and often even undergoing disfiguring and life-changing surgery. Here I was symptom free, and I had a strong desire to share my experience on the off chance that anyone else might then manage to achieve the same result that I had, yet no one else had heard of it.

My research also found that Symprove was a relatively small private company but its star was clearly rising. In July 2020 they concluded a deal with a fantastic investment partner when newly formed London-based private equity group, bd-capital, announced an investment. bd-capital's press release for the deal highlighted many of the points made in this chapter:

> Interest in how to improve gut health has grown exponentially in recent years, as advances in scientific research and medicine have continued to uncover how the diversity and balance of bacteria in the microbiome affects many conditions – not only gut conditions such as IBS and IBD, but Parkinson's and dementia, cardiovascular conditions, mood and mental health. While the science is emerging, it has become clear that people with many different diseases tend to have a less diverse or unbalanced microbiome.

A key point of differentiation for Symprove is that it seems able to deliver sufficient quantities of live and active bacteria to the gut. 'Good bacteria' found in fermented foods and in many probiotic products often don't make it past the stomach due to how acidic the stomach is. As Symprove explains:

> Digestion involves a lot of seriously strong stomach acid, triggered by the ingestion of food. Bacteria delivered in food are exposed to these harsh acidic conditions and less likely to survive. [...] Because Symprove is water-based, it doesn't trigger digestion. More live and active bacteria surviving passage through the stomach means a better opportunity to colonise the gut.

More studies are needed in humans to provide concrete evidence that the product has clinical efficacy across various health problems. This having been said, based on the evidence of many thousands of users of the product and on the opinion of a number of key clinicians and scientists, it seems highly likely that the product does have efficacy for IBS and IBD sufferers at the very least. The placebo effect may have a role to play here, but I suspect that whether or not that is the case is of little consequence for the many users of the product who have experienced 'life-changing' results, me included.

This story is an example of the new sorts of treatment modalities which may hold the key to many of the health problems we confront nowadays, and which may have come as a result of the changes we have made to the microbiome in and around us in the decades since we have triumphed over infectious disease.

Symprove is by no means the only company working in this area. My hope is that there will be a great deal more focus on dysbiosis as a potential key causal factor in the rise of so many of the most significant health problems prevalent nowadays.

As we have already seen, dysbiosis has been implicated in a wide range of health problems. As one study by Wing Yin Cheng and colleagues published in *Gut* has put it: '[D]ysbiosis [...] contributes to the development of various pathological conditions, including obesity, diabetes, neurodegenerative diseases and cancers.' And as another 2019 study published by Fattorusso et al., has said about the potential link with autism specifically: 'In recent years, there has been an emerging interest in the possible role of the gut microbiota as a co-factor in the development of autism spectrum disorders (ASDs), as many studies have highlighted the bidirectional communication between the gut and brain (the so-called "gut-brain axis").'

I don't mean to suggest that dealing with dysbiosis, potentially by using effective probiotics, is 'the' answer to the contemporary global obesity or mental health epidemics, or a solution to all of the diseases of modernity, but it seems like it may be a pretty good place to start and certainly merits a great deal more investigation and focus.

At the very least, this broad approach should be given serious consideration by medical professionals. At the moment this happens all too rarely, for one specific and, to my mind, rather worrying reason: amazingly enough, all over the world, the vast majority of our clinicians receive little or no training in nutrition. As Michael Greger MD puts it in the introduction to his book *How Not to Die*:

> Our diet is the number-one cause of premature death and the number-one cause of disability. Surely, diet must also be the number-one thing taught in medical schools, right? Sadly, it's not. According to the most recent national survey, only a quarter of medical schools offer a single course in nutrition.

Even that minority of medical schools in the United States which do offer some training in nutrition give it little focus.

Much the same is true in the UK, and I have first-hand experience of this reality. When I was first diagnosed with ulcerative colitis my GP told me that the condition was almost certainly nothing to do with diet. Seen through the prism of what I have learned since then this does seem to me one of the more ridiculous things that a medical professional has ever said to me, but it is perhaps not that surprising when you consider that in the UK, a recent study of 853 medical students and doctors found that more than 70 per cent had received fewer than two hours' nutrition training while at medical school.

That is fewer than two hours in five or six years of study and is also the reason I've met several GPs who were not familiar with the term 'dysbiosis'. As a BBC news article by Sheila Dillon put it in 2018: 'We learn nothing about nutrition, claim medical students.' Happily, the direction of travel is at least reasonably encouraging. As an example, in reaction to this state of affairs, UK University of Bristol medical students Iain Broadley and Ally Jaffee set up Nutritank in 2017, to promote 'the need for greater nutrition and lifestyle medicine education within healthcare training'. Today Nutritank is franchised in more than 25 British medical schools, was recognized by the UK Prime Minister's 'Points of Light' awards, and is now contributing to the NHS long-term plan in the UK National Health Service.

It is precisely these sorts of grassroots initiatives which I believe will make a difference over time. Let us hope that they do.

Micronutrients

Another piece of the puzzle concerns micronutrients: vitamins and minerals. Dr Rhonda Patrick is a US research scientist who regularly talks about the crucial role played by micronutrients in our health. She has explained the concept of 'micronutrient *inadequacy*' rather than 'deficiency'. As she puts it: 'Inadequacy means you're not so deficient that you're going to have a clinical symptom that's measurable like your gums are falling out. But you will, for example, not be making collagen properly.'

Dr Patrick's broad thesis is that our modern diets leave us with micronutrient *inadequacy* across a wide range of key vitamins and minerals which are vital for our health. These are not severe enough to be picked up by clinicians in our 'sick care' system as deficiencies which cause a serious health problem in the short term, but they will likely increase the chance that you suffer from chronic disease if your micronutrient consumption is inadequate for many years as is the case for so many people. These inadequacies are then very likely compounded by dysbiosis given how important our microbiome is for the effective absorption of micronutrients.

By now, millions of us, in the developed world at least, intuitively understand the importance of micronutrients in our diet and the global supplement industry is worth tens of billions of dollars as a result. What is reasonably clear, however, is that there remain question marks over the quality and bioactivity of many of the products which are sold and their efficacy as a result.

What also seems more than likely is that the microbiome has a key role to play. If your gut flora is compromised, your ability to assimilate even top-quality supplements will be challenged.

Intermittent fasting

Another key related idea advocated by Dr Patrick and others like her is that of intermittent fasting, or 'IF' for short. This is a dietary approach which has gained significant attention in recent years, with proponents suggesting a number of key health benefits. Most obviously, fasting can lead to weight loss by reducing overall calorie intake. Arguably more important, however, is the role it can play when it comes to insulin sensitivity and blood sugar control, and in cellular repair and metabolic health overall.

Fasting triggers a process called *autophagy*, which involves the breakdown and recycling of damaged cells and cellular components and the removal of toxins. Recent research suggests that the practice can improve heart health, reduce inflammation, improve cognitive function and guard against neurodegenerative diseases such as Alzheimer's and Parkinson's. Interestingly, intermittent fasting seems likely to have a positive impact on the gut microbiome, improving the diversity and composition of gut flora and an individual's microbial resilience.

To increase the chance of an optimal health outcome, the key seems to be to ensure all of the bases are covered – a good-quality diet in the right quantities, the cultivation of a healthy microbiome, and effective supplementation to ensure micronutrient adequacy if needed. These constitute three legs of a stool and are all interrelated such that the removal of any one of them will increase the risk of illness, obesity or mental health problems and seriously compromise an individual's likely 'healthspan' – the period of their life where they will be largely free from illness. Too many of us are missing at least one leg of the stool in seeking optimal health.

Oxygen – 'the forgotten nutrient'?

In fact, there is yet another important fourth leg to that stool which is all too frequently forgotten by medical practitioners and their patients alike, in much of the world at least: the role played by our breath and breathing.

Oxygen has been described by research papers and nutritionists alike as 'the forgotten nutrient'. Because we don't eat it, we tend to forget about it when implementing a diet programme or just more generally when it comes to optimizing our health overall. We spend a great deal of time thinking about protein, fat, carbohydrates, vitamins and minerals. If we are lucky and taking account of the arguments made in this section of the book, we may also think about the composition of our gut flora, but all too seldom do we explicitly factor our breathing habits into the equation.

This is highly problematic given what a crucial role it has to play in so many of our biological processes. This reality is brought home more or less robustly when you consider that going without oxygen for just a few minutes will result in nothing short of death. The same cannot be said about protein, fat, vitamin C or even water, which is next in line after oxygen in terms of keeping us alive.

Yet notwithstanding this reasonably dramatic reality, too few of us think about our breathing sufficiently frequently or give it any kind of concerted focus. Because breathing happens automatically, most of us don't spend any time explicitly thinking about it or seeking to improve or even optimize our oxygen 'consumption' regularly throughout the course of our lives.

Of course, one of the key benefits of exercise is that it increases our oxygen intake. Any moderate or strenuous exercise will increase our heart rate and the rate at which we respire which is one of the key reasons that it is a good idea. But I would argue that few of us are explicitly thinking about this when we exercise, and even if we exercise regularly, the rest of the time we will more than likely revert to breathing on autopilot rather than proactively seeking to improve our breathing.

This may be another reason why so many chronic diseases are caused by modernity. Numerous studies have found that our hunter-gatherer forebears were significantly more active than most of us are nowadays. Hunters and subsistence farmers walked tens of thousands of steps a day and ran far more often than most of us in the developed world do today. This may also go some way to explaining why so many of these diseases are vastly more prevalent in the developed world than in the developing world. Even today, people like the Maasai in Kenya and Hadza in Tanzania, for example, are far more physically active than most people in the West. This is another factor which may explain the health outcomes we have already looked at for the Maasai over and above their diets, for example.

A few pages ago we looked at growing evidence for the likely benefit of a regular meditation practice. One of the key features of meditation is a focus on breathing. Anyone who spends even a few minutes a day meditating will, by definition, be spending those few minutes focused on their breathing. This is likely to be a more or less prosaic explanation for why meditation seems to confer positive health outcomes (although my personal belief is that those benefits very likely go some way beyond breathing alone).

Taken together, someone who exercises and meditates regularly will be consuming more oxygen in any given week than someone who does neither. Over time, it seems likely that this may result in better health outcomes for that individual. But even if someone meditates and exercises daily this will only constitute a few hours a week where they're optimizing or, at the very least, improving their oxygen intake. The vast majority of the rest of the week they will be breathing on autopilot which may mean they're not getting as much oxygen as might be optimal, particularly if they are stressed which tends to result in shallow breathing and is a condition which is all too prevalent in our modern lives.

A concerted focus on breathing as often as possible in any given day is almost certainly a good thing for many of us to contemplate, particularly those of us suffering poor health of one kind or another. One of the top-selling non-fiction books of 2020 was James Nestor's *Breath: The New Science of a Lost Art.* It is a fascinating look at the science behind breathing, at just how important it is for so many elements of our health, at how bad we are at doing it and how little emphasis we place on it nowadays, especially when compared to how it was viewed by many cultures throughout history. As Nestor puts it:

> Modern research in pulmonology, psychology, biochemistry, and human physiology is showing us that making even slight adjustments to the way we inhale and exhale can help jump-start athletic performance, rejuvenate internal organs, halt snoring, allergies, asthma, and some autoimmune disease, and even straighten spines. None of this should be possible, and yet it is.

In the book Nestor provides compelling evidence for the fact that human breathing patterns have changed dramatically over the past few thousand years and makes the case for the benefits of certain ancestral breathing practices, and particularly the importance of nasal breathing as against breathing through our mouths.

At some level we are all familiar with this when a well-meaning friend or relative suggest we should 'take a deep breath' at a time of stress, for example. Nestor's book goes into much greater detail and reveals just how much more there is to it. It makes a compelling case for how singularly important breathing is overall and for the nuanced role of carbon dioxide as well as oxygen.

Wim Hof – breathing and hormetic stress...

Throughout his book Nestor describes several 'pulmonauts' – pioneers in our understanding of breath and breathing. One such figure is Wim Hof. Wim is a wonderfully charismatic Dutchman, nicknamed 'the Ice Man' who has set more than two dozen Guinness World Records. He has completed a marathon in the Arctic Circle barefoot and wearing only shorts, hung on one finger at an altitude of 2,000 meters suspended from an air balloon, run a marathon in the Namibian desert without drinking any water, and spent fully 112 minutes in a container full of ice – one of his world records. Crucially, he has also proven scientifically that the autonomous nervous system related to the innate immune response can be purposely influenced.

His book *The Wim Hof Method* has been an international bestseller, he has a huge following all over the world, and counts an extraordinary roster of A-list celebrities as advocates and clients, including the likes of Chris Hemsworth, Orlando Bloom, Sacha Baron Cohen and Oprah Winfrey. He has also been the subject of a BBC television-reality show *Freeze the Fear*, although don't let any of this put you off!

Perhaps more interesting than the Hollywood actors and television personalities who are effusive fans are some of the world-leading athletes who credit his methods with significantly improving their performance. These include Laird Hamilton, one of the greatest big-wave surfers in history who co-invented 'tow-in' surfing and surfed the 'heaviest' (that is, quite possibly the most dangerous) wave in the world at Teahupo'o, French Polynesia, and Dutch martial artist Alistair Oversteem, one of only two fighters to hold world titles in both mixed martial arts and K-1 kickboxing simultaneously. The list of such supporters is long.

Wim Hof's 'method' rests on three key pillars: breathing, cold exposure and what he calls 'commitment'. That last 'pillar' is really just about ensuring that you add each of the first two to your habits with consistency. Just as with any habit or behaviour, real benefit invariably accrues over time. There is little point adding breathing exercises and cold exposure to your life if you only do so for a week or two and then stop.

Over time, however, Wim and the scientists who have been studying him for some years are beginning to demonstrate the merits of a concerted focus on breathing and on the considered use of cold exposure too.

The first pillar of the 'method' is a set of simple breathing exercises, performed at least once a day but ideally more than that, after lunch to combat the post-lunch slump that so many of us suffer perhaps, or before a particularly demanding meeting, presentation, or sports fixture for example. As I suggested above, implementing a daily habit of focused breathing can result from any number of practices, including meditation and yoga, or simply as a result of taking regular exercise, but this does seem to be an area where, broadly, 'more is better' so having another 'call to action' to focus on your breathing is quite likely to be a good idea.

Cold exposure

Equally interesting is Wim's advocacy of regular cold exposure. This can range from the simple practice of adding a minute or two of cold to the end of your daily shower; through cryotherapy, which has become extremely popular in many professional sports; all the way to rather more radical activities such as taking ice baths, swimming in frozen lakes and, for real devotees, taking walks in freezing conditions wearing not much more than shorts and a T-shirt.

The key underlying idea here is that of 'hormesis' or 'hormetic stress'. As Wim explains, hormesis:

> describes a phenomenon in which a substance or environmental agent known to be harmful in larger doses has stimulating and beneficial effects on living organisms when the quantity of the harmful substance is small. Living cells actually adapt in response to these substances (or stressors), positively affecting their condition and functionality.

Most of us are broadly familiar with this idea when it comes to exercise. Any worthwhile gym regime should seek to exert the right amount of hormetic stress on our muscles and/or cardiovascular system. Too little hormetic stress in our lives and we will gradually sink towards obesity, lethargy, and, in the long run, very likely a raft of health problems and mental health problems. Too much and we may cause ourselves real harm and injury. The right amount, however, will make us bigger, stronger and fitter and enhance our health and happiness across the piece.

What seems to be key when it comes to cold exposure is that it is a hormetic stressor for our *vasculature*. It has been estimated that our vascular system of arteries, veins and capillaries is more than 60,000 miles long. Most of us are already more than aware that it is utterly crucial for our health given that it is responsible for providing our cells with oxygen and carrying away carbon dioxide. It is instrumental for many of our other biological processes including our ability to assimilate nutrients, for example.

It would seem, therefore, that anything that might improve the effective functioning of those 60,000 miles of critical plumbing might have merit. More research needs to be conducted but the evidence does seem to be more or less clear by now that cold exposure has precisely that effect. Cold exposure challenges our body to open up our vascular system in a bid to warm us up. Again, most of us are more than familiar with this reality – this is one of the reasons our cheeks turn red when we are cold, for example. When challenged regularly over time, however, this function will improve as a result. This is no different from what happens to our muscles when we perform a bicep curl, squat or bench press in the gym.

Nowadays, our warmly clothed and overly comfortable lives spent sheltered indoors have meant that our vascular system isn't anywhere near as challenged as it was in the past and as it was designed to be when many human beings spent far more time being cold, and this is increasingly being implicated in all sorts of health problems. To combat this reality, we need to train our vascular system just as we do our muscles. Happily, this is relatively simple and may only need us to add a minute or three of cold-water exposure to our shower each morning.

Heat exposure

Interestingly, there is a fair bit of evidence that heat exposure has similar merit for similar reasons. Earlier in the book I made reference to the prevalence of sauna-bathing in many times, places and cultures throughout the world, whether in Turkey, Japan or Scandinavia for example. Taking a sauna is also a hormetic stressor. On her website Dr Rhonda Patrick marshals the increasing amount of evidence that shows that sauna use can increase lifespan and improve overall health. One study she points to, focused on more than 2,300 middle-aged men from eastern Finland, 'identified strong links between sauna use and reduced death and disease', notably cardiovascular conditions but also others such as dementia and Alzheimer's disease:

> [M]en who used the sauna two to three times per week were 27 percent less likely to die from cardiovascular-related causes than men who didn't use the sauna. [...] Men who used the sauna roughly twice as often, about four to seven times per week, experienced roughly twice the benefit – and were 50 percent less likely to die from cardiovascular-related causes. In addition, frequent sauna users were found to be 40 percent less likely to die from all causes of premature death. These findings held true even when considering [... several other] factors that might have influenced the men's health.

It seems that regular cold exposure can deliver similar outcomes. This position is being taken increasingly seriously by academics and clinicians alike. Earlier in the book I made reference to *The Telomere Effect*, a superb book which was written by Dr Elissa Epel and Dr Elizabeth Blackburn. Dr Blackburn won the Nobel Prize in Medicine in 2009 for her work on telomeres. I find it interesting, therefore, that the forward to Wim Hof's book was contributed by none other than Dr Epel. Quite clearly these ideas are being taken seriously by serious people.

Throughout Wim's book there are individual testimonials from people sharing their own positive and often moving experiences of the impact of having used breathing and cold exposure to combat: arthritis, ulcerative colitis, multiple sclerosis, breast cancer, chronic pain, bipolar disorder, depression, obesity, and stress. In the preface to his book, Wim himself states without any moderating or qualifying language: 'People who have embraced my method have been able to reverse diabetes, relieve the debilitating symptoms of Parkinson's disease, rheumatoid arthritis, and multiple sclerosis; and address a host of other autoimmune illnesses, from lupus to Lyme disease.'

These are obviously bold claims and certainly open him to accusations of quackery and potentially of running ahead of the science, but my own view is that we should almost certainly give him the benefit of the doubt. To date he has shown himself more than willing to participate in scientifically rigorous studies to examine the efficacy of his claims and even to proactively cultivate interest from scientists and clinicians. He is constantly advocating and pushing for more research.

I find these ideas compelling if only given how logical they seem. It certainly seems highly unlikely to me that there is much if any downside in improving our habits both around breathing and challenging our vascular system with moderate cold and/or heat exposure.

Even if the effusive testimonials of thousands of people who claim to have enjoyed meaningfully positive health outcomes from using such techniques turn out to be a function of the placebo effect alone, frankly who cares? If they are achieving those outcomes, what does it matter if all that is involved is a cold shower and slightly more focus on breathing each day?

My own personal experience of such things is that you feel great in the moment when you do them anyway. I have been enjoying euphoria-inducing cold exposure since childhood when I used to swim in the North Atlantic off the coast of Northern Ireland (where my father is from). I have also enjoyed regular saunas whenever possible over the years in a succession of overpriced gyms and business hotels.

My own view is that the science will more than likely support the merits of regular breathing exercises and cold (and hot) exposure for a broad range of health conditions, just as it will when it comes to the role of the microbiome and dysbiosis, and we probably don't have that long to wait before it does. To a certain extent these habits have been verified scientifically already, but more studies are needed to push things to the next level and to drive considerably more uptake of such ideas in our healthcare systems.

As an example, a May 2018 study published in the *Journal of Applied Physiology* made reference to the apparent therapeutic potential of cold exposure. As its authors, Yoanna M. Ivanova and Denis P. Blondin, put it: 'In healthy individuals, cold exposure can increase energy expenditure and whole-body glucose and fatty acid utilization. Repeated exposures can lower fasting glucose and insulin levels and improve dietary fatty acid handling, even in healthy individuals.'

The study went on to note the fact that 'for centuries, there have been claims and anecdotal evidence extolling the virtues of cold exposure' and then went on to conclude: 'Further work is needed to determine the frequency, minimal intensity, duration, and type of cold exposure required to elicit meaningful metabolic changes in individuals.'

My belief is that the research needed here will be delivered and there is a reasonable chance that healthcare systems may adopt such practices over time, even if, sadly, this takes several years longer than would have been optimal as is so often the case.

Sleep

Another key factor for our health which is increasingly well understood is the role played by sleep. I think it is fair to say that by now most of us have a reasonable understanding of the importance of sleep for our health and wellbeing. That said, as with so many such things it seems that a large percentage of the population gives it insufficient focus, particularly in the developed world. We have a very serious sleep problem. In his 2017 book *Why We Sleep: The New Science of Sleep and Dreams*, leading neuroscientist Matthew Walker provides an in-depth exploration of the importance of sleep. As he explains at the start of the book: 'Two thirds of adults throughout all developed nations fail to obtain the recommended eight hours of nightly sleep.' This is extremely problematic for the health and mental health of many millions of us and for society more generally. It is no coincidence that the inexorable rise of so many of the diseases of modernity we have looked at over the last many pages is most prevalent in those nations where sleep time and quality has declined most dramatically over the last few decades: the United States, much of Western Europe and other modern developed nations such as Japan, South Korea, Australia and New Zealand. This is yet another key risk factor of modernity and affluence.

In terms of our physical health, sleep is mission-critical for repairing and restoring our body and particularly important for the effective functioning of our immune system, our cardiovascular function and for our entire metabolic system overall. As a result, chronic sleep deprivation is associated with a higher risk of obesity, diabetes, cardiovascular disease and a raft of other serious health problems. Over time, poor sleep habits can double your risk of cancer. Lack of sleep also impairs cognitive function, leading to poor decision-making and an increased probability of accidents.

Just as important in our stressed-out modern world, where depression and many other mental illnesses are so clearly on the rise, sleep is essential for our *mental* health. Sleep regulates our moods and emotions and is instrumental for the effective functioning of our key cognitive processes. Chronic sleep deprivation leads to depression, anxiety, suicidal ideation and many other mental health problems.

More generally, good sleep is also crucial for learning and memory. During sleep, the brain consolidates new information and strengthens memory. Lack of sleep can significantly impair memory and learning. As a result, poor sleep can have seriously negative consequences for our academic trajectory at school and university and on throughout our career.

One of the most important factors in good sleep is 'regularity' – how consistent we are in the time we go to bed each night and get up each morning. The less regular our sleep patterns, the more likely we are to suffer from insomnia and the lower the quality of our sleep overall. For many decades now, particularly in modern developed nations where we can afford such things, it is considered entirely normal behaviour for so many of us, particularly teenagers and young adults, to massively impact the regularity of our sleep several times a week. Whether this be all-night partying on a Friday or Saturday night or watching television or playing video games in to the 'wee small hours', such behaviour is considered entirely normal by many millions of us.

If someone in their teens or twenties (or thirties, forties, fifties or sixties for that matter) needs to wake up at, say, 6.00am from Monday to Friday each week for school or for their job, but goes to bed at, say, 2am or even 4am on Saturday and/or Sunday morning many weekends of the year, they are subjecting themselves to a weekly 'jetlag' effect roughly equivalent of a multi-thousand mile flight to the other side of the world. This wreaks havoc with our all-important circadian rhythms and, by extension, our sleep, health and mental health.

This was not the case in most times and places throughout human history. The vast majority of our species had no access to an inexpensive light source after dark until the last century or so. This meant that most things and most people shut down once night fell. This is still the case in much of the developing world, particularly in agrarian communities who still have no electricity. Many such communities also simply don't have the economic wherewithal to enjoy all night parties or similar on any given weekend. Might this be yet another explanation for the materially lower instance of mental health problems in such communities, even given how relatively poorer they are than those of us in the developed world?

As trite as it sounds, many of our health and mental health problems may have a great deal to do with cultural habits which are impacting our sleep, particularly when it comes to our teenagers and young adults where the problem is compounded further by the significant stressors of study, taking their first tentative steps in the labour market and finding a significant other.

As one example, attention deficit hyperactivity disorder (ADHD) has risen inexorably in recent decades in much of the developed world. According to the US Centers for Disease Control and Prevention, the prevalence of ADHD among US adolescents (ages 12–17) had risen to fully 13 per cent by 2019. That is around 3.3 million American teenagers. There is a growing body of literature that suggests that sleep problems may be a significant contributor to ADHD, and that improving sleep may lead to improvements in ADHD symptoms. The same can be said of so many other physical and mental health problems.

Rather than prescribing pills to treat such things, thoughtful clinicians and scientists such as Matthew Walker are increasingly advocating more of a focus on poor sleep as a key causal factor and better sleep as a healthier and cheaper potential treatment. When you consider that many people suffering from ADHD and so many similar problems may also be suffering from dysbiosis and, indeed, that poor sleep and a poor microbiome are also almost certainly related, might the answer to so many of our health challenges lie somewhere other than several million Western teenagers being prescribed drugs such as Ritalin and Adderall and several million more adults anti-depressants?

I don't want to sound like some draconian, curmudgeonly party-pooper here, in terms of suggesting that teenagers and twenty-somethings should all be tucked up in bed by 9pm seven days a week, but it is perhaps worth highlighting the extent to which we have normalized behaviours which are certainly having an exceptionally deleterious impact on the health and happiness of millions of us in the developed world, and from an early age. Such habits are then carried on into later adulthood and are highly problematic for us as individuals and for society as a whole.

Many of our solutions to such problems seem to miss key underlying root causes so often hiding in plain sight, and all too often our healthcare systems resort to the sticking-plaster treatment of symptoms with drugs. Happily, however, there are reasons to believe that things may improve somewhat in future.

There is increasing evidence that young adults in the developed world today are more health-focused than was the case in the past. For some years now, there has been increasing interest in physical fitness and 'wellness' overall, which is particularly pronounced in young adults. In the USA gym membership of 18–34 year olds has increased by more than 44 per cent since 2012, for example.

Young adults in the developed world are becoming more health-conscious in their food choices and with regards to their mental health, with increased attention on the importance of self-care, mindfulness and stress reduction. There is also far less of a stigma associated with seeking mental health support than in the past. Fewer young adults are choosing to smoke or use drugs than in the past, and rates of alcohol consumption are also falling. This all bodes well for the future trend in our quality of sleep and for a great deal else besides.

These trends are due to increased education and awareness of the negative health effects of these behaviours. The internet and social media have provided young adults with access to vastly more information than was the case in previous generations and a wealth of health-, diet-, sleep- and fitness-related apps. In many of the wealthiest places in the world, social media influencers promoting healthy living and improved diets are the 'rockstars' of the era. Much of this information can be a bit hit and miss, but these are likely to be better role models in terms of health outcomes and social cohesion than the chain-smoking, booze- and drug-addled rockstars of the last few generations, throwing televisions out of hotel rooms!

Again, the (bio)tech industry has a role to play here. First, the industry is helping us to better understand the importance of sleep and to provide evidence for the causal relationship between sleep and so many of our health and mental health problems. (Bio)tech companies are giving us the information we need to better understand such things, working to deliver tools which can help us improve our behaviours, and the nutraceuticals and pharmaceuticals which can also improve matters.

Such tools include sleep tracking apps and wearable devices. I have been using the Sleep Cycle app on my iPhone since 2010, for example. There is no question that it has had a fantastic impact on my quality of sleep over more than a decade, but there are far superior technologies emerging, particularly when you consider innovative products such as the Apple Watch and Oura Ring.

As the evidence-based ideas of scientists such as Matthew Walker gain more traction, dare we hope that the experience of the next few decades might be to reverse the deterioration in our sleep that has reached epidemic proportions in the last few?

Exercise and 'movement'

We have already mentioned the importance of regular exercise in passing in the context of improving our breathing habits, but it is worth taking a moment to look at the subject in a little more detail.

Dr Peter Attia is an award-winning medical doctor whose health, performance and longevity-focused podcast *The Drive* is one of the most highly rated in the world. He is regularly on record making the case that the two single most important factors for our health and longevity are 'VO2 max and muscular strength'. He explains that these two factors are 'in a league of their own' and, amazingly enough, that the positive impact of having a high

VO2 max (which measures the maximum amount of oxygen your body can use during exercise), and being physically strong can outweigh many of the things that we might think of as having a negative impact, including smoking, high blood pressure and even diabetes. As he puts it in an interview with Tim Ferriss: 'The downside of those things is relatively small compared to the upside of having a high VO2 max and being very strong.'

As Attia explains, someone moving just from the bottom quartile of the population in terms of VO2 max to the third quartile cuts their risk of all-cause mortality in half. This is an outcome which no drug or vitamin regime can deliver. A higher VO2 max improves cardiovascular health, materially reduces the risk of chronic disease and improves health and longevity as a result.

We improve our VO2 max with aerobic exercise, which is more or less well understood by most of us, but also with more of a concerted focus on breathing, underscoring again the ideas of people like James Nestor and Wim Hof we covered a few pages back.

Attia also stresses the crucial importance of building *physical strength*, however, ideally through resistance training – lifting weights. This has a raft of health benefits over and above cardio-focused exercise alone, and helps to prevent age-related muscle loss and maintain bone density, all of which is highly correlated with longevity and increased healthspan.

A key point here, which is made by Attia and many others like him, is that too few of us focus on strength training and resistance weights when contemplating exercise, either sufficiently or at all. An increasing number of people are doing a good job when it comes to improving their VO2 max with aerobic exercise, but far fewer place any concerted emphasis on resistance training, which is likely to be prejudicial to their potential long-run health outcomes.

A third crucial piece of the puzzle concerns the importance of movement and mobility over and above aerobic fitness and physical strength alone. Kelly and Juliet Starrett are a high-profile couple working in the health and fitness world and well known for their work in this area. Kelly has consulted with top athletes from the NFL, NBA, NHL, US Olympic team and both the UK's Premier soccer and rugby leagues as well as all branches of the US elite armed forces. Juliet is a former professional white-water rafter, who won five US national titles and three world championships in that discipline.

In their book *Built to Move*, they make the case that the human body is designed to move far more often and through a far wider range of motion than is the case for most of the human population today. Movement is fundamental to human health and longevity. It is essential for proper joint function, the efficient functioning of many of our most important physical processes such as our digestive, endocrine, lymphatic and respiratory systems, and for preventing injury, particularly as we age. It helps us maintain physical strength, agility and balance. Crucially, it also plays a key role in mental health and wellbeing – all of which reduces the risk of chronic disease and increases healthspan.

What I find particularly eye opening about their work is the idea that many of us may think we are doing a good job with physical exercise, if we undertake several vigorous gym sessions each week for example, but that a great deal of that good work can be undone if we are highly immobile the rest of the time.

If you do 30 or even 60 minutes of strenuous work on a stationary bike or treadmill early in the morning, for example, but then habitually spend the next 12 or 14 hours sitting at an office desk or slumped on a sofa watching television, you won't be deriving nearly as much benefit from that exercise as you might hope and you may still be heading inexorably towards chronic health problems and a heightened risk of injury and back pain in particular as you get older.

This point is particularly well made when you consider the extent to which so many of us who exercise tend to pick one exercise in particular, or relatively few, to the exclusion of everything else and perform that exercise repetitively over many years, even decades. This is a point made by the Starretts and others like them: in childhood, most of us are very active. Importantly we tend to be so across a wide range of activities and a wide range of movements as a result. Small children and school children climb trees, leap from bed to bed, sit on the floor cross-legged, ride their bikes, sprint fairly often in playground games and tend to engage in a wide range of organized activities and sports, often including team and individual sports and swimming for example.

As we age, our range of activities tends to narrow significantly. With the time-pressures of college or university and then on to work and parenting, even those of us who do our best to keep up with an exercise regime of some kind will tend to morph from doing a number of different sports or activities in our childhood and teens to one main activity or very few. While it is certainly better to be doing any exercise as against none at all, being particularly focused only on one activity to the exclusion of all others can be problematic for health and longevity and increase the risk of injury, particularly over long periods of time.

Happily there are a number of relatively simple steps we can take to combat this reality. We need only ensure we incorporate some key movements and habits into our daily routine to significantly improve our mobility and range of movement. There is mounting evidence that doing so can be highly beneficial and is some way more important than many of us realize.

One of the key 'steps', in the most literal sense, is to increase the time we spend walking each day. The Starretts stress that walking is 'intrinsically tied to the robustness of all the systems and structures in your body' and go on to say that: 'The simple act of walking surpasses any fitness gadget or club membership you can buy.' They also make the point that someone who does an hour of strenuous aerobic exercise in the morning and spends the rest of the day almost entirely sedentary will be significantly worse off than another individual who does the same morning session and then also walks a few thousand steps later in the day.

The idea that we should all aspire to walking 10,000 steps a day has been around since the 1960s. It was originally the brain child of a Japanese pedometer company whose primary focus was, perhaps understandably, on selling pedometers. The idea came in for a measure of criticism in the years that followed with some suggesting it wasn't based on scientific research and others concerned the idea was too simplistic to have merit. Critics argued that the 'right' number of steps for optimal health for a given individual would vary enormously based on any number of factors including their age and general level of fitness.

More recently, however, the consensus is increasingly that said Japanese pedometer company may have been fairly prescient, even if they were running ahead of robust scientific evidence at the time. The Starretts reference a large 2020 study which found that 'compared with taking 4,000 steps per day, reaching 8,000 steps was associated with a 51 per cent lower risk of death from all causes. Taking 12,000 steps per day was associated with a 65 per cent lower risk.'

From a health and longevity perspective, walking several thousand steps a day is an unalloyed good. Doing so will improve your cardiovascular and respiratory health, your bone density and the functioning of your immune system, thereby reducing your risk of infection and other illnesses, certain cancers, obesity, diabetes, osteoporosis and a great deal else besides. Regular walking has also been shown to have a significant positive impact on mental health outcomes, reducing the risk of depression and anxiety.

The word 'regular' is important here. Walking two miles every day of the week is fundamentally better for us than doing nothing all week and a 15-mile hike on Saturday and Sunday – although walking two miles a day *and* doing the hike is best of all, of course.

A reasonably large percentage of the population takes far too little exercise or even does no exercise at all. In recent years, however, the evidence is showing us that even those that do may be insufficiently focused on the crucial importance of strength-training and mobility and movement, both of which are likely to be more important than we realized in the past. Happily, such things are relatively easy to add to your regime.

In summary

Again, the biotech industry as I'm defining it is giving us the tools to understand all of the above in much greater detail and the information and apps to ensure we might apply that knowledge consistently. In future, the industry is also best placed to deliver genetic and other analyses which will help us understand what the best approach to such things may be for us individually – in terms of diet, sleep and exercise, for example.

Biotech companies can and will also provide nutraceuticals and even pharmaceuticals that might help us optimize our approach to how we exercise by improving our energy levels, for example, or super-charge the progress we might make if we have neglected exercise in the past and need to get back on the proverbial horse as quickly as possible.

In the next, and final, chapter of this part of the book, we will look at the key role played by our expectations and mindset, at the merit of 'little and often' and developing long-duration habits when it comes to our health and at how the biotech industry can help with all of the above as it moves us towards the idea of 'medicine 3.0'.

7
The importance of expectations and of 'little and often'

Earlier in the book we looked at increasing evidence for and acceptance of a so-called 'mind–body' link. This is another key area where I believe great progress is likely going forwards. It is almost a decade since I read Deb Shapiro's eye-opening book *Your Body Speaks Your Mind: Understand the Link between Your Emotions and Your Illness.* As she explains in the first chapter of the book:

> During the past ten years there has been a growing body of research showing how the mind and body respond to each other, clearly demonstrating how emotional and psychological states translate into altered responses in the chemical balance of the body. This in turn affects the immune, neural, endocrine, digestive, and circulatory systems.

Shapiro's book was originally published in 2007, so she is here talking about the decade from 1997 onwards in terms of that 'growing body of research'. Things have moved on a great deal since then with far more work having been done by scientists all over the world.

Expectations shape experience

David Robson is a British science writer who is a former features editor at the *New Scientist.* In his 2022 book, *The Expectation Effect: How Your Mindset Can Transform Your Life*, he 'takes us on a tour of the cutting-edge research happening right now that suggests our expectations shape our experience'. As he puts it: 'People who believe ageing brings wisdom live longer. Lucky charms really do improve an athlete's performance. Taking a placebo, even when you know it is a placebo, can still improve your health.' I don't think it is overly controversial to suggest that many of these sorts of ideas were traditionally viewed with some cynicism by the medical profession and by the general population as a whole. For many years, the notion that you might be able simply to use 'positivity' and 'think' your way to better health, fitness or mental health was treated with a good measure of scepticism. It still is today by many.

The mood music is changing here, however, as we increasingly come to understand and evidence that your psychology *can* significantly impact your physiology, particularly over long periods of time. These ideas have actually been hiding in plain sight for several centuries: the Greek Stoic philosopher Epictetus, who died in 135 CE, famously said, 'People aren't disturbed by things, but by the views they take of them'; Shakespeare has Hamlet say, 'There is nothing either good or bad, but thinking makes it so'; while, somewhat closer to our own time, President Abraham Lincoln is reputed to have said, 'Most folks are about as happy as they make up their minds to be.' If the same basic idea seems to be shared across about 20 centuries by no less consequential a set of human beings than one of the truly great ancient Greek philosophers, the world's most famous playwright, and conceivably the United States' greatest president, there is probably some merit in their position.

The nocebo effect

This belief in the power of positive expectations is particularly important when we consider the impact of another idea covered in Robson's book: that of the 'nocebo effect'.

The nocebo effect is effectively the opposite of the placebo effect. If the placebo effect involves a positive expectation that leads to a positive outcome, the nocebo effect involves a negative expectation that leads to a negative outcome. This can be as important to our health and happiness as the placebo effect can be, and shows that the mind–body connection can work both ways. Negative expectations can have a powerful impact on our health. Robson cites studies which have shown that people who expect to experience negative side effects from a treatment or medication are more likely to actually experience those side effects. Amazingly enough, this can even happen if the treatment in question is itself a placebo!

People who expect a medication to be ineffective may actually experience reduced benefits, even if the medication has proven to be highly effective in a clinical trial. People who think of themselves as being fundamentally sick or who constantly talk about or dwell on a given health problem are more likely to be sick and/or suffer from that health problem.

The nocebo effect can also be influenced by how treatments are framed and presented to clinicians and patients alike, and by cultural and social expectations more broadly, which can mean that more inherently pessimistic cultures may derive less benefit from a given drug or treatment. As a UK national, I do have to wonder whether the curmudgeonly British national character may have a role to play here, although I concede I've found no scientific basis for that position!

The nocebo effect means that addressing negative expectations head-on can improve outcomes. There is an increasing focus on such things from the pharmaceutical industry and medical community, with input from specialist psychologists. By addressing and changing negative expectations, healthcare providers can potentially improve treatment outcomes and reduce negative side effects.

This is all fairly powerful stuff. Happily, it does seem that such ideas are becoming a great deal more popular and far more widely accepted. There is growing interest in the mind–body relationship from research scientists and the general public alike. We can see this when we consider the hundreds of millions of consumers who have downloaded apps that are explicitly focused on the mind–body relationship including the likes of *Calm* and *Headspace* I have already referenced, for example.

Little and often

Over the last several pages of the book, we have looked at a number of factors which are fundamental to our health, several of which I would argue have been given insufficient focus by clinicians and by society as a whole until really very recently. Some of these ideas are gaining traction, but their adoption by healthcare systems all over the world remains extremely patchy, even if the direction of travel seems mildly encouraging. These factors have included: the role of the microbiome over and above a focus on diet and calorie consumption alone, the problem of micronutrient 'inadequacy' (rather than deficiency), and the importance of oxygen consumption, hormetic stress, sleep, exercise, and our psychology and mental attitude for long-run health outcomes.

I think it is worth explicitly drawing out a unifying theme when it comes to addressing all of these factors: the crucial importance of time and consistency. Anyone suffering from dysbiosis, for example, will very likely require several months to repair the damage done to their microbiome over many years up to that point. Addressing dysbiosis is not something that can be achieved overnight or even over several days. Much the same can be said about any of the other factors listed above. To enjoy the full impact of improved breathing, exercise or sleep habits or of the impact of causing hormetic stress to our vasculature takes quite some time – months, years or even decades.

A lifetime of commitment to such things and the implementation of the idea of 'little and often' is key to achieving the optimal outcome. We all intuitively 'know' that 'good' daily or weekly habits will have a positive effect on us over time, whether those habits are to do with work, finance, fitness or health, but too many of us very significantly underestimate the sheer *scale* of that effect over long periods of time.

As an illustration, consider the impact of an individual committing to two simple daily habits from the age of 20 onwards: ten push-ups and a brisk two-kilometre walk. A 60-year-old who has implemented only those two simple daily habits (of push-ups and brisk walks) over 40 years will have done nearly 150,000 more push-ups and walked nearly 30,000 kilometres more than a 60-year-old who has developed no such habit. All other things being equal, of course, it seems as if this will more than likely impact their relative levels of health and fitness. Crucially, you may not see much difference between those two individuals at thirty or even forty, but you almost certainly will by the time they both hit sixty and beyond.

This is particularly the case when you consider that anyone implementing those sorts of habits also more than likely improves them over time. Ten push-ups a day becomes twenty and more over the course of several years as the habit becomes deeply ingrained and strength improves.

James Clear is one of the world's leading experts on building good habits. His book *Atomic Habits*, first published in 2018, has sold more than 15 million copies since its publication. The book is an excellent articulation of just how powerful such small daily actions can be for our lives. As Clear puts it: 'Tiny changes, remarkable results.'

Importantly, I would argue that this broad concept is of real relevance to the various ideas covered in this part of the book. 'Little and often' over many years, if successfully applied to our diets, the composition of our microbiome, our stress levels, our breathing habits and so much else besides, will very likely hold the key to delivering significantly better health outcomes than we are currently experiencing and, by extension, better and more enjoyable lives for so many of us.

The fact that these results will only come optimally over many years of consistent habit is also one of the reasons that our healthcare systems so often struggle to deliver the outcomes we need, and why so many of these diseases of modernity are on the rise as we have seen.

To a great extent, in most of the world we have a short-term 'sick care' system which attempts to treat symptoms, rather than a long-term *health*care system which addresses underlying causes. By now this idea has certainly entered the contemporary zeitgeist to judge by how often such arguments are made in the press and by politicians, scientists, entrepreneurs and medics alike. We certainly need a 'sick care' system and, as we have already seen, I think it is worth dwelling on just how far we have come in that respect in the last century or more and, perhaps, being more grateful for that progress than many of us are.

That said, if we are to significantly improve matters going forwards, seek to address the inexorable rise of so many of these diseases of modernity, improve the health of the general population overall, and, indeed, ensure we stand ready to combat any future pandemics that might arise, we can probably do better.

Medicine 3.0

This idea of 'little and often' is a key tenet of what is increasingly being referred to by leading scientists and clinicians as the idea of 'medicine 3.0'.[1] 'Medicine 3.0' is being used to describe a potential future for medicine that is far more focused on 'healthcare' throughout our entire lives, as against our current system of 'sick care' when something goes wrong, by which time it can often be too late for much more than treating symptoms, sticking-plaster solutions and expensive drugs which might increase lifespan but often don't deliver much in terms of healthspan.

The successful delivery of this new and improved approach to medicine overall rests on the development of several key emerging technologies, most of which will be delivered by the biotech industry. These include personalized or precision medicine where treatments or nutritional programmes can be tailored to an individual based on their unique genetic make-up,

---

1. Medicine 3.0 stands in contrast to 'medicine 1.0', which refers to traditional or conventional medicine and 'medicine 2.0', which refers to the transition of paper-based systems to the use of digital technology and the internet in healthcare.

lifestyle, age and other factors, and the use of advanced diagnostic technologies and digital technologies such as wearable devices, telemedicine, and health apps to monitor and manage health, as well as the application of artificial intelligence and machine learning to all of the above.

The fundamental idea behind medicine 3.0 is to create a more holistic, patient-centred, data- and technology-driven approach which can keep an individual healthy throughout their lives and potentially reduce or retard the need for 'medicine 1.0'-type treatments or eliminate them altogether. Such an approach stands to significantly improve outcomes for many of us as individuals and also, in the fullness of time, for healthcare systems overall by improving efficiency and significantly reducing costs.

The role of biotech in medicine 3.0

Crucially, for the purposes of this book, I believe that the (bio)technology industry can have a key role to play here. There are a number of reasons for this. First, there is extraordinary complexity involved in working out the science that underpins all of the ideas shared so far in this section of the book. There are quite literally millions, if not billions, of variables when we contemplate the diversity of human, bacterial and viral DNA in and around each of us, for example, and then we need to layer on top of that the enormous variability in our individual diets, habits and environments. It is very hard to control for all of that – with our current approaches and tools at least.

By now it is pretty clear that dysbiosis more than likely has a key role to play when it comes to many of our most intractable healthcare problems and 'modern plagues'. A great deal more research is required, however, to hone our understanding of dysbiosis, how it impacts those healthcare problems and what we should be doing about it. At present our knowledge is pretty imperfect, our diagnostic and analytical tools are pretty blunt and certainly not fully formed, and healthcare systems are not addressing dysbiosis in any particularly concerted way as a result.

It is the biotech industry which is more or less obviously best placed to deal with all of the above. It is biotech companies which can and will conduct the research required to move our understanding of such things forward. It is life sciences companies that will develop the tools which enable us to diagnose dysbiosis far more accurately than in the past. Biotech companies will also more than likely deliver the new therapeutic treatments and nutraceuticals which can address it and apps which ensure compliance with a given treatment plan over relatively long periods of time. There is also even a chance that such therapeutics and habit-forming apps will be highly targeted and personalized to someone's individual genome, biome and health condition in the not-too-distant future.

Similarly, armed with the insight that there is no such thing as a one-size-fits-all 'good diet' and the understanding that the 'right' diet for each of us individually will be a function of the extremely complex interplay of our own genome, biome, virome and environment, it is innovative life sciences companies that will help us work out what the best diet for each of us is individually, and also very likely develop therapeutic probiotics that will sit alongside a diet plan and fast-track a patient's recovery from dysbiosis. This could be a more powerful approach to treating all sorts of health problems than those we currently employ – particularly when it comes to obesity, for example, where we know that dieting and calorific restriction alone rarely work, in any durable way at least (see Chapter 6).

To a certain extent this is happening already. There are companies which offer home-based DNA testing, the largest and best known of which is 23andMe. The company is based in California and has genetically tested more than 13 million customers since its foundation in 2006. As the company explains: 'We're all about real science, real data and genetic insights that can help make it easier for you to take action on your health.'

This approach is a good example of a company working on the distinction between sick care and healthcare, the merits of 'little and often', and engendering behavioural change over long periods of time. The company reports that 76 per cent of its customers make healthier choices, 55 per cent improve their diets, 51 per cent set explicit health goals and 45 per cent exercise more. As a result, 23andMe likes to argue in its investor presentations that it is to healthcare what YouTube has been to media, Uber to transportation and Airbnb to hospitality – a genuine disrupter. Not only can it exert a long-run influence on its patients' health choices based on rigorous science but it has also been able to build an enormous database of personalized (and anonymized) healthcare information. The company already has more than 4 billion data-points which it can use internally for its own therapeutic programmes, or in partnership with biopharma companies to assist in drug development and quite possibly more personalized therapies in the fullness of time. It has a collaboration with British pharmaceutical company GSK in this area already, and announced the extension of that deal in January 2022.

23andMe is certainly an interesting and innovative company but has 'only' tested just over 13 million people in the developed world to date. This is clearly something of a drop in the ocean in the context of a global population of 8 billion of us. It is also focused only on human DNA, which, as we have seen, may prove to be a relatively small piece of the puzzle years from now as we develop a deeper understanding of the crucial role played by microbes and microbial DNA and also of something called the 'epigenome' – how that DNA is 'expressed' in our cells.

There will be many more companies working in this area in future with increasingly sophisticated tools and technologies. 23andMe may enjoy some kind of first mover advantage over time, not least given the size of its database, but this is an area where many of the biggest and best-funded companies in the world are committing billions of dollars of capital and hiring the brightest and best, including Apple, Google/Alphabet, Microsoft and many others. This is also an area where innovation can and will come from companies which are either very small today or which don't even exist yet given the exponential pace at which this sort of science can develop.

Innovative biotech companies will also no doubt continue to interrogate the merits of breathing exercises, meditation, hormetic stress from cold (or heat) exposure, micronutrient inadequacy, the importance of movement and strength, the expectation effect, and a great deal more over time. It is biotech researchers who will further prove out the ideas of James Nestor, Wim Hof, Dr Rhonda Patrick, Matthew Walker, Peter Attia, Kelly and Juliet Starrett, David Robson, and an army of others like them, and provide the tools required for us to embed such ideas and practices into our lives and, hopefully, our healthcare systems too in the fullness of time.

Without as well as within

Arguably even more exciting, in the long run, such companies will very likely be able to address problems 'without' us as well as 'within'. As I have already suggested at various points in the book, these technologies may well be able to play an exciting role in very significantly improving the world around us over and above simply working on therapeutics. Biotech companies can roll back environmental degradation, give us new and significantly cleaner ways of producing the food and energy we require, and even ensure that Moore's law (see Chapter 2) continues into the future by giving us new technologies to improve processing power as we begin to hit the limits of physics with respect to chip design.

Of course, if we improve our environment, we will very likely improve our health too, given how many of our 'modern plagues' seem to be related to the impact we have had on such things since the Industrial Revolution. Our health and the 'health' of our environment are more or less obviously related.

This is particularly likely to be the case as we improve our understanding of the key role played by microbes in all of the above – not just within us as we have already examined, but without us too. As Martin Blaser highlights in his book *Missing Microbes*: 'We live on a microbial planet that is totally dominated by forms of life too small to be seen by the naked eye. [...] They make the oxygen we breathe, the soils we till, the food webs that support our oceans.' Even with a population of around 8 billion of us, it has been estimated that human beings make up only approximately 0.1 per cent of all the biomass on earth. Bacteria, on the other hand, comprise fully 12.8 per cent. People sum to only 0.06 gigatonnes of the biomass of life. Bacteria total more than 70. They are more than *one thousand times* more consequential in terms of life on earth than we are, and this is even before we consider fungi and archaea which comprise another 19 gigatonnes.

It seems increasingly likely that many of the biggest problems we currently confront as a species may have quite a bit to do with our impact on that microbial biomass, particularly since the Agricultural and Industrial Revolutions, and their impact on us in turn. It stands to reason, therefore, that many of our solutions may come as a natural function of a far better understanding of the microbial world, and it is precisely this which so many biotech companies and research scientists are seeking to do.

Crucially, we are only just beginning the journey towards improving our understanding of such things. It is only in the relatively recent past that we have even developed the tools to examine the microbial world in any meaningful way and only in the last few years that we may even have developed the tools to make changes to it. Importantly, however, it is also very likely that such things will develop exponentially, just as they have with most other new scientific frontiers of human discovery in the time since the Industrial Revolution.

Real wealth creation and widespread adoption

Over and above delivering the science which enables us to understand all this complexity, and the new therapeutics and tools and technologies which may improve our health and our environment, the other key role which I believe the biotech industry has to play will have to do, rather more prosaically, with the creation of significant real wealth.

In dealing with all of the above and delivering an explosion in new technology across so many key areas of human development to help solve so many of our most 'valuable' problems, the industry will deliver several trillion of real economic value in the decades ahead just as the tech industry has done in the last few.

The output of all that wealth creation will be to increase human living standards still further. It will do this *directly* given the value created by all those new tools, systems, techniques and treatments, and *indirectly* simply by the overall economic contribution of the industry to financial markets and government coffers, for example. Many of those new tools and treatments will be far more widely available than may seem feasible at present, and they could be widely adopted as the industry works to bring cost down, just as has happened in so many other industries in the last several decades. We will surprise ourselves with just how quickly the technology comes together in this area and then becomes commonplace.

We will have revolutionary new treatment modalities which will fast-track our recovery from any number of the key healthcare challenges we face at present. At the same time, there will also be a much wider adoption of smart technologies which help us personalize and improve every aspect of our health throughout our lives.

It won't be too long in the future that we may have extremely sophisticated AI personal assistants embedded in our smartphones or whatever else may replace them, in combination with some kind of wearable or even implantable device. These technologies will be able to tailor our habits and behaviours and how we eat and exercise throughout our lives in real time, such that healthcare outcomes are significantly improved and, more generally, we are able to lead happier, more energetic, productive, longer and less stressful lives overall as a result.

Such technologies will move us away from sick care and towards true healthcare. To a certain extent the wealthiest citizens in the world are already some way along on this journey, if you consider the behaviour of early adopters in places like Silicon Valley and things like the 'quantified self' movement, but biotech innovation will also mean that the cost of all of the above comes down enough for such practices to become far more widely adopted across the world.

Just as smartphone adoption has outstripped all but the very most bullish estimates, largely as a result of cost reductions achieved for those products, the same will very likely happen with many of the new technologies I believe will arrive in the years ahead, and this will likely pay enormous dividends for our species and for our planet.

We will also be able to work with the microbes which so dominate our planet to achieve a wide range of beneficial outcomes for our environment and for nature. To some, this kind of concerted intervention in the natural world may seem like a worrying case of 'playing God', particularly when we consider that we will have the ability, increasingly, to genetically engineer the DNA of living things. My reading suggests, however, that progress will proceed sufficiently carefully and with involvement from all the key stakeholders required, including scientists, clinicians, regulators, governments, and even philosophers, journalists and religious leaders, for this concern to prove to be overblown. We will return to this subject when we look at CRISPR and gene editing towards the end of the book.

The phrase 'if man were meant to fly, God would have given him wings' is reasonably well known, yet today our species takes several billion plane journeys per annum and that particular behaviour is broadly accepted, even celebrated. In the years ahead, much the same will happen with many of the beneficial innovations that will come to us from the biotech industry even if some of them quite rightly cause us a measure of soul-searching in how we go about developing and using them initially.

If scientists can show that the use of many of these technologies will actually be returning our microbial world to 'how it used to be' before the impact of our industry altered it to our detriment as a species, this may go some way to allaying those concerns about 'playing God'.

In summary

In this part of the book, we have looked at how modern medicine has given us the tools to combat infectious disease, yet has also, in turn, unleashed so many chronic 'diseases of modernity'.

We have examined the role played by the microbial world and dysbiosis as a key causal factor for so many of those health conditions. To address such things most effectively, and improve our health more generally, we need to continue to improve our understanding of the complex interplay between our diets, lifestyle choices, habits and our very own genetic make-up and microbiome.

It is the biotech industry which is giving us the science required to understand this complexity and the tools and treatments required to make a tangible difference to the trajectory of these many chronic health conditions.

In the next and final part of the book, we will look in more detail at a number of the key technologies being developed by the industry.

PART 3

The past, present and future of biotech

In this part of the book, we look at how the biotech industry has developed, from the early days of the pharmaceutical industry, and on through the discovery of DNA, RNA and mRNA, and at some of the pioneering scientists who have paved the way for today's breath-taking advances.

We then look at some of the most cutting-edge technologies being delivered by the industry, including gene therapy, stem cells and gene editing and at the health economics of such treatments.

We examine the expansion of our diagnostic and analytical tools: PCR, NGS, and the increasing role played by artificial intelligence and machine learning, and consider the role biotech has to play 'without us' – in areas such as clean power generation, agriculture and bioremediation, and even computing. While these topics may seem tangential to human health and wellbeing, they are, of course, fundamental – human beings can flourish only on a planet that flourishes after all.

In the final chapter, we examine the undeniably radical idea that age may be nothing more than a disease, and one that may even prove curable in time, and what this could mean for us as a species.

All of the above is likely to have a profound and positive impact on us in the years ahead.

8
The development of the industry

I finished Part Two suggesting we might now turn our attention to some of the extraordinary technological advances which could give us hope that the future may be brighter than so many of us fear. I'm going to kick things off by looking at therapeutics (drug development). This is the area of progress which could bring us effective cures for cancer and all sorts of other terrible diseases and be instrumental in driving several trillion or more of real wealth creation as a result after all.

Begin at the beginning

Probably the first point to make about drug development historically is just how extraordinarily hit and miss it was. Early 'drugs', such as they existed, tended to come about as a result of observation and quite often serendipity or blind luck. They were also invariably natural extracts of one kind or another.

One of the earliest such examples was the use of the bark of the *Cinchona* or 'quina-quina' tree to derive quinine for the treatment of malaria and tropical fevers. In the 1600s Jesuit missionaries in Latin America observed that the bark was used by Indigenous peoples in the Andean jungle to treat fevers and brought it back to Spain, hence its nickname 'Jesuit's bark'. More than a century later the Swedish founder of modern taxonomy, Carl Linnaeus, named the plant in honour of the Spanish Countess of Chinchón, wife of the Viceroy of Peru, who, according to legend, brought the bark back to Spain in the 1630s after she herself had suffered from a fever which had been cured by the bark.

For the first two centuries or more of its use, cinchona bark was first dried, powdered and then mixed into a liquid as a drink. It wasn't until the first half of the nineteenth century that two French chemists, Pierre-Joseph Pelletier (1788–1842) and Joseph-Bienaimé Caventou (1795–1877), were able to isolate quinine and several other 'alkaloids'. In his book *Cracking the Code*, British investor Jim Mellon describes the resulting compound – quinine sulphate – as 'the first mass-produced drug'. This marked a crucial turning point in therapeutics, which began to shift away from the preparation of crude plant extracts to the chemical formulation of natural and synthetic compounds.

The shift was accompanied by a move towards the industrial production of drugs. This was a key step as you might imagine. The foundations for mass production had been laid nearly two centuries earlier when there was a gradual move away from small, local apothecaries towards larger laboratories, even if these were still focused on natural products.

The German company Merck is usually credited with being the oldest true pharmaceutical company in the world. Founded as long ago as 1668, it was arguably the first to become a research-based industrial company from about 1827 onwards when it began to offer a 'collection of high-purity alkaloids' to chemists, physicians and pharmacists.

Over the next several decades things developed at an increasingly rapid pace, propelled by advances in the production of synthetic dyes. As Jim Mellon puts it: 'Much of the progress at the time was derived from a single invention that was made by an 18-year-old English student, William Perkin, who combined nitric acid and benzene [...] to create the first artificial dye'. The invention of dyes made the world a fundamentally more colourful place but, perhaps even more importantly, it was soon discovered that 'drugs derived from the advances in dyestuffs could successfully be used to treat infectious diseases'.

This insight came in large part as a result of work done by the German scientist Paul Ehrlich (1854–1915), one of the founding fathers of the pharmaceutical industry and winner of the 1908 Nobel Prize in Physiology or Medicine. Working with dyes, Ehrlich and his Japanese assistant, Sahachiro Hata (1873–1938), discovered the compound Arsphenamine, otherwise known as Salvarsan, primarily used as a treatment for syphilis. Salvarsan is generally credited as being 'the first anti-infective small molecule'. According to Jim Mellon, Salvarsan marked the introduction of synthetic chemicals in treating disease. Dr. Ehrlich's work laid the groundwork for many practices still observed in pharmaceutical research today. It was the first instance of systematically evaluating multiple compounds in laboratory settings.

It was this *systematic* approach which was key. This set the stage for the development of many other 'small molecule' drugs in the decades that followed and even up to the present day where such drugs still comprise an estimated 90 per cent of all pharmaceutical drugs.

The German company Hoechst manufactured Salvarsan at scale. Hoechst began life as a dye manufacturer which gave it the chemistry skills and manufacturing facilities required for early pharmaceutical production. The same could be said of yet another German company, Bayer. Bayer had synthesized aspirin a few years earlier in 1899. Not long after Ehrlich's discovery of Salvarsan in 1910, Bayer followed up with Luminal (phenobarbital) for epilepsy in 1912, Germanin (suramin), an early antiparasitic used to treat sleeping sickness and river blindness in 1923, and Prontosil, an early antibacterial drug in 1935. Thousands more 'small molecule' drugs have followed in the century or so since then.

Penicillin and anti-infectives

The next really consequential step forward came in the 1940s with the invention of penicillin. Notwithstanding the fact that Dr Ehrlich and many others like him had systematized pharmaceutical research, the discovery of penicillin gives us another example of the key role so often played by serendipity and observation in drug discovery.

In the summer of 1928 the Scottish physician and microbiologist Alexander Fleming – later to become Sir Alexander Fleming – went on holiday to his country house in Suffolk. On his return to his laboratory at the beginning of September he found that the Petri dishes on his bench had gone mouldy. In arguably one of the most fortuitous chance discoveries in history, he noticed that the mould had killed off the bacteria that he had been growing. The specific bacteria was *Staphylococcus aureus*, a bacterium that can cause skin infections. He found the mould to belong to the genus of fungi known as *Penicillium*. Happily, in subsequent experiments, Fleming established that *Penicillium* had the same effect on several other disease-causing bacteria. He called the 'mould juice' that he managed to extract from the bacterium 'penicillin'.

It took more than a decade before the Oxford chemists Ernst Chain (1906–79) and Howard Florey (1898–1968) worked out how to isolate and purify enough penicillin for clinical use in May 1940, and several more years before the team managed to convince the US government to provide the funding and enough manufacturing capacity to produce the drug at scale. By the end of the Second World War, however, penicillin was seen as something of a miracle drug and carried by Allied military units all over the world.

Fleming, Chain and Florey were awarded the Nobel Prize in Physiology or Medicine in 1945 'for the discovery of penicillin and its curative effect in various infectious diseases'. Arguably, the most important outcome of that discovery was the development thereafter of the entire class of treatments we call anti-infectives. Since the invention of penicillin scientists have developed dozens of other antibiotic and anti-infective treatments which have played a critical role in combatting disease and infection for decades.

Edward Jenner and 'the speckled monster'

Chance and observation also had a role to play in the development of vaccines. Until its eradication in 1980, smallpox had been one of the most devastating diseases facing our species throughout history. Some historians cite it as a contributory factor to nothing less than the decline of the Roman Empire given the millions of Roman citizens killed by the Antonine Plague from 165 CE onwards. It was also responsible for many million more deaths across the 'New World', as the Spanish, Portuguese, British and French carried it to the Indigenous peoples of the Americas. It is hard to understate just how impactful and devastating the disease was. In his seminal 1997 book, *Guns, Germs, and Steel*, US scholar Jared Diamond suggests that smallpox and other infectious diseases may have killed fully 95 per cent of the Indigenous population across the American continent.

Even notwithstanding the fact of the COVID pandemic of recent years, from the vantage point of today, it is hard for us to comprehend just how frightening and ferocious a disease smallpox was to our forebears. As a 2005 study by Dr Stefan Riedel puts it:

> Smallpox affected all levels of society. In the 18th century in Europe, 400,000 people died annually of smallpox, and one third of the survivors went blind. The symptoms of smallpox, or the 'speckled monster' as it was known in 18th-century England, appeared suddenly and [...] were devastating. The case-fatality rate varied from 20 per cent to 60 per cent and [...] in infants was even higher, approaching [...] 98 per cent in Berlin during the late 1800s.

If COVID had inflicted that kind of fatality rate, we would have just lived through a period of somewhere between 1.2 billion and 4.8 billion deaths globally and there could now be as many as a billion more blind and horribly disfigured individuals in the world.

Given just what a hideous disease smallpox was, understandably there was a good deal of focus on working out how to prevent it all over the world. In places as diverse as China, India, Africa and Turkey and as long as several centuries ago, people noticed that survivors of the disease became immune to it thereafter. The idea then developed that, if you could infect a healthy person with a mild dose of the disease from a sufferer, that healthy individual might then develop immunity.

There is documented evidence for the adoption of this broad approach from all over the world. As long ago as the fifteenth century, members of the Chinese elite were drying scabs taken from sufferers, powdering those scabs and blowing that powder up a patient's nose. Elsewhere, people would purchase scabs or even pus from an individual suffering with smallpox and then rub that material into a small cut made somewhere on their own skin. This practice became known as 'variolation'. As gruesome and basic as these practices may seem to us, they did have the benefit of working a fair bit of the time and were widely adopted as a result.

The British aristocrat and writer Lady Mary Wortley Montagu (1689–1762) is often credited with popularizing the practice in the UK from the early 1720s onwards. She was the wife of a British diplomat in Constantinople (present-day Istanbul) and had noticed that the practice of variolation was widely spread in Turkey. Having suffered the disease herself, and lost her brother to it, she asked Charles Maitland (c.1668–1748), a Scottish surgeon also based in Constantinople, to variolate her children. Her decision led a member of the Royal Society in London, the physician James Jurin (1684–1750), to publish a survey comparing the mortality rates from natural smallpox to the milder infection caused by variolation. His work and that conducted by other scientists and physicians led to the gradual acceptance of the practice in the UK and beyond during the course of the eighteenth century.

Although variolation was a significant improvement over the status quo, it had drawbacks. As Riedel's study explains: '2 per cent to 3 per cent of variolated persons died from the disease, became the source of another epidemic, or suffered from diseases (e.g., tuberculosis and syphilis) transmitted by the procedure itself.' One particularly high-profile fatality from the process was that of Prince Octavius, the youngest son of King George

III who died in May of 1783 not long after having undergone variolation against smallpox. It was against that backdrop that the English surgeon and physician Edward Jenner (1749–1823) introduced the world to a new and improved approach to the prevention of smallpox: vaccination.

Born in Gloucestershire, England, Jenner finished his medical studies at St George's Hospital London in 1770 before moving back to his home town of Berkeley to practise medicine from 1772 onwards. Variolation was part of his practice, just as it was for any trained doctor at the time, but he was more than aware of its limitations. Living in a rural community, Jenner was, according to the Jenner Institute's website, 'intrigued by country-lore which said that people who caught cowpox from their cows could not catch smallpox'. Cowpox was a much milder disease which caused smallpox like 'pocks' to appear on their udders. It was common for milkmaids in rural communities to catch the disease from their animals. As the Jenner Institute explains: 'Although they felt rather off-colour for a few days and developed a small number of pocks, usually on the hand, the disease did not trouble them.' This was obviously a rather less drastic outcome than catching smallpox. Jenner wondered whether mild cowpox could confer the same immunity on patients as using the far more dangerous smallpox used in variolation.

From 1796 onwards he conducted a number of experiments to test his thesis and published his findings in 1798. The new approach was called 'vaccination' which was derived from the Latin word for cow, *vacca*.

Although the significance of Jenner's work was recognized reasonably quickly, with the British government awarding him large grants in 1802 and 1807, there was widespread resistance to this new practice of vaccination for several decades. Some opposed the use of a product derived from an animal on religious grounds. More prosaically, plenty of doctors were averse to the technique because it threatened their lucrative variolation businesses. These are themes which we looked at earlier in the book when we considered just how long it can take before a new treatment achieves widespread adoption (Chapter 2). In the case of Jenner's vaccine, it took the best part of half a century. An Act of Parliament outlawed variolation in 1840, and vaccination using cowpox was enshrined in British law in 1853. The point is often made that Jenner wasn't the first person to come up with the idea of vaccination. An honourable mention should be made of another British physician, Benjamin Jesty (c.1736–1816), who used cowpox more than two decades before Jenner, but it was Jenner's tireless advocacy for the practice which lead to its widespread, if belated, acceptance and adoption.

It is hard to calculate the tangible impact of that work in terms of however many hundreds of millions or even billions of lives have been saved and misery averted since then, but as the Jenner Institute has put it: 'It has been estimated that the task he started has led to the saving of more human lives than the work of any other person.' A key point to stress here is that it took Jenner and his collaborators nearly half a century to achieve broad acceptance for vaccination. Happily, it took only around ten months for today's vaccine scientists to gain broad acceptance for a COVID vaccine, even if that work rested on several decades of research up to that point. If that kind of exponential development continues into the future, one day we might be able to engineer effective vaccines and other medicines in hours, minutes or even seconds, as fundamentally extraordinary as that might sound from the vantage point of today.

Small and large molecules

Over the last few pages we have looked at how 'small molecule' drugs and vaccines developed and seen that small molecules make up the significant majority of approved drug treatments. It is perhaps worth taking a moment to clarify what we mean by 'small molecule' drugs, if only to contrast them with a new class of drugs that have emerged in the last few decades and which can broadly be described as 'large molecules'.

Small molecules are defined as those organic compounds with low molecular weight. A dalton is a unit of measurement for molecular weight usually calibrated against the mass of a carbon or hydrogen atom, and small molecule drugs are conventionally considered to be any molecule which is less than about 800–1,000 daltons (that's roughly equivalent to 800–1,000 hydrogen atoms).

As we have already seen, chemists have been able to synthesize small molecules and manufacture them at scale for more than a century. Small molecules have a number of features which make them suitable for use as therapeutics. First, they are 'relatively' simple. As a result, they are cost-effective to manufacture and can be produced reliably at scale. Many of them are also stable, meaning that a small molecule drug can often be made into tablets and kept at room temperature for long periods of time without degrading, which keeps supply chain costs down. This stands in contrast to more complicated drugs which may need to be shipped in liquid form or even at very low temperatures throughout the supply chain and which may have a much shorter shelf life.

The molecular structure of small molecules also means that they can pass through the membranes of human cells to reach 'intracellular targets'. In plain English, they can get into our cells and 'do something'. They can also be engineered so that they can be taken orally in tablet form as an inactive 'prodrug' and then gradually *metabolize* into an active drug as they move through the body and react to what they find there, generally our bodily fluids. Pharmaceutical companies have been fine-tuning such things for more than a century now and it is pretty incredible what they can do with the chemistry in this respect. Tablets can be coated and a molecule engineered so that the active ingredient is released only when it arrives in a certain part of the body for example.

From the 1980s onwards, however, an entirely new class of drug began to emerge: 'large molecules'. As the name suggests, large molecules are sophisticated therapeutics which are significantly larger and more complicated than their small molecule counterparts. Specifically, they tend to have a molecular weight of between about 5,000–50,000 daltons in size (5–50 kilodaltons, or kDa).

Large molecule treatments are usually built from proteins or peptides (chains of amino acids), some of the most fundamental building blocks of life. These are obtained from biological sources using complicated biotechnological manufacturing processes, and such drugs are therefore also generally referred to as 'biopharmaceuticals' or 'biologics'.

Although, as we have already seen, small molecule drugs still make up the lion's share of therapeutics by *number*, large molecules constitute a significant and growing percentage of drug treatments by *value*. By 2020, seven of the top ten best-selling drugs in the world were biologic drugs adding up to many tens of billions of dollars' worth of revenue. This is partly because they are significantly more expensive to manufacture than small molecule drugs, for now at least, but also because they have shown efficacy treating diseases which were previously largely or entirely untreatable, in areas such as diabetes and cancer for example. This has all happened reasonably quickly. Although the first biologic drug emerged as long ago as 1982 when Eli Lilly launched a drug called Humulin for diabetes – the first commercially available biosynthetic human insulin – many of today's most valuable biopharmaceuticals were approved only in the last few years.

Large molecule drugs are another one of the emerging technologies which underpin the fundamental thesis of this book given their impact on certain diseases, particularly cancer, the value they have created to date and, even more exciting, what they may be able to deliver going forwards.

To credit that position, it is worth looking at the scientific progress that was required before scientists would be able to synthesize such exceptionally complicated molecules.

DNA – the discovery of the 'double helix'

One of the most important forward steps taken on the road to the development of large molecule therapeutics, and to a great deal else besides, was Francis Crick (1916–2004), James Watson (1928–) and Maurice Wilkins's (1916–2004) discovery of the structure of DNA (deoxyribonucleic acid) in the early 1950s (for which they won the Nobel Prize in 1962). As the US National Library of Medicine has put it:

> In short order, their discovery yielded ground-breaking insights into the genetic code and protein synthesis. During the 1970s and 1980s, it helped to produce new and powerful scientific techniques, specifically recombinant DNA research, genetic engineering, rapid gene sequencing, and monoclonal antibodies, techniques on which today's multi-billion dollar biotechnology industry is founded.

As is essentially always the case with such discoveries, their work rested on a great deal of work done by other scientists. To quote another scientific titan, Sir Isaac Newton, Watson, Crick and Wilkins were certainly 'standing on the shoulders of giants'.

As long ago as 1869, the Swiss chemist Friedrich Miescher (1844–95) was actually the first scientist to identify a substance which he called 'nuclein'. This was later renamed 'nucleic acid', which is the 'NA' in both DNA and RNA (ribonucleic acid). Nucleic acid was nothing less important than the fourth fundamental building block of life, with the others being proteins, lipids and carbohydrates.

Despite being little known, Miescher's work was a crucial early first step on the road to understanding DNA. Another player who was much closer in time to Watson, Crick and Wilkins was Erwin Chargaff (1905–2002). Working at Columbia Medical school in New York from 1935 onwards, he completed important work on the ratio of the key components contained within the structure of DNA: guanine, cytosine, adenine and thymine. Chargaff even met with Watson and Crick in 1952, and his work is widely understood to have made a key contribution to theirs, something he would write about later given his disappointment at having been excluded from consideration for their 1962 Nobel Prize.

Similarly, work completed at around the same time by the American chemist Linus Pauling (1901–94) – the only person in history to have been awarded two unshared Nobel Prizes in two different fields – provided a critical foundation for what came afterwards. In particular, he discovered the single-stranded alpha helix (the structure found in many proteins) and led the way in developing the method of model building in chemistry which Watson and Crick were to use in uncovering the structure of DNA.

Perhaps the foremost scientist who deserved rather more credit for her contribution seen from the vantage point of today was the young X-ray crystallographer Rosalind Franklin (1920–58). Her crystallographic 'photographs' of the structure of DNA were instrumental in Watson and Crick's discovery and that work was also the reason that her colleague Maurice Wilkins was the third person named on the prize. Sadly for Franklin, she died in 1958, from ovarian cancer that was more than likely caused by her many years of working with X-rays. This was four years before the 1962 Prize was awarded, and although there was not yet a rule against Nobel Prizes being awarded posthumously at the time, it was generally not done.

Rosalind Franklin's expertise lay, primarily, in the use of X-ray crystallography. This was an extraordinarily innovative technology that had developed only from 1912 onwards. It was a crucial innovation in that it enabled scientists to analyse materials down to the level of atoms and chemical bonds for the first time. X-ray crystallography was then also used by Franklin and others, including Linus Pauling, to determine the structure of many biological molecules, including DNA of course. Similarly, in his work, Erwin Chargaff relied on the use of paper chromatography and ultraviolet spectroscopy, both of which techniques had been developed only in the early 1940s.

High-profile big leaps forward such as Watson, Crick and Wilkins's work on DNA were only possible thanks to the progress being made by thousands of scientists all over the world, and with the development of dozens of 'revolutionary' new scientific techniques which were complementary and self-reinforcing. Such progress was also being made in an increasingly globalized world where such techniques, ideas and technologies were shared like never before. This phenomenon was arguably particularly pronounced after the end of the Second World War. Not only were scientists no longer working in secrecy and isolation for their respective warring nation states, but a large number of them were able to leave a shattered Europe and Japan for a USA which could provide them with ample funding, a highly developed system of patent law and the very latest tools and equipment. This was supercharged in the decades which followed by the arrival and spread of air travel and the development of computers and the internet of course.

From Hippocrates, Dioscorides and Galen, to Edward Jenner and Lady Montagu and on through Paul Ehrlich, Alexander Fleming, Watson, Crick and Franklin and all the way up to the Jennifer Doudnas of today, progress has rested on an extraordinarily complex and self-reinforcing series of forward steps scientifically, culturally and financially. From a slow start two thousand years ago, such things have developed at an increasingly exponential pace.

In terms of biotech specifically and of the development of large molecule biopharmaceuticals, the elucidation of the structure of DNA fired the gun on several decades of increasingly rapid progress which has already delivered some incredibly beneficial science and resultant therapeutics and diagnostics and which is getting us closer than ever to the very real possibility of an effective cure for cancer and any number of other terrible diseases and health problems.

From DNA to RNA and mRNA

As the 1950s proceeded, Watson, Crick and plenty of others like them continued their work on DNA and on the other key nucleic acid, RNA. Over time they established the relationship between DNA, RNA and proteins and, by extension, nothing less consequential than how life is 'built'.

In 1958 Francis Crick published his 'central dogma of molecular biology' which explained that DNA makes RNA and that RNA in turn makes proteins. As science writer Ruairi Mackenzie explains in a December 2020 article for *Technology Networks*: 'DNA replicates and stores genetic information. It is a blueprint for all genetic information contained within an organism' and 'RNA converts the genetic information contained within DNA to a format used to build proteins.' As he puts it with a wonderfully helpful and modern analogy: 'DNA is a storage device, a biological flash drive that allows the blueprint of life to be passed between generations. RNA functions as the reader that decodes this flash drive.'

In the decades that followed Crick's articulation of that 'central dogma', scientists increasingly came to understand the detail. In 1961 two articles appeared in *Nature*, announcing the isolation of messenger RNA (mRNA). Messenger RNA copies the snippets of genetic code which can be seen, say Damian Garde and Jonathan Saltzman in a special report for *STAT*, as 'a recipe book for the body's trillions of cells'. Given its role, '[f]or decades, scientists have dreamed about the seemingly endless possibilities of custom-made messenger RNA, or mRNA. [...] The concept: By making precise tweaks to synthetic mRNA and injecting people with it, any cell in the body could be transformed into an on-demand drug factory.'

Perhaps the most familiar real-world use of this technology at the time of writing is the fact that two of the three main COVID vaccines developed to combat the pandemic are mRNA vaccines: those developed by Moderna, whose very name is a portmanteau of 'modern' and 'RNA', and by BioNTech together with Pfizer.

In the last 60 years or so, dozens of scientists have been awarded Nobel Prizes for work related to the understanding of RNA. A key point to make here is that, even if those two mRNA vaccines were approved in record time, it took more than half a century and upwards of 30 Nobel Prizes to progress from the isolation of mRNA to that real-world use therapeutically. In fact, one of the most critical developments to get the technology over the line was delivered reasonably recently by two scientists, Katalin Karikó (1955–) and Drew Weissman (1959–), who had laboured in relative obscurity and without the support of their colleagues, institutions and grant-funding bodies for many years in order to provide a final crucial piece of the mRNA puzzle.

Karikó and Weissman's story would make for a good Hollywood movie and one will probably be made in the fullness of time, but their key contribution was to figure out how to stop the body's immune system's response to synthetic DNA. For decades, the promise of any kind of RNA-derived treatment ran into the problem that the body's immune system would destroy any RNA introduced for a therapeutic purpose because it would see that newly delivered genetic material as 'foreign' or 'alien'. Worse, that immune response also increased the risk of a therapy causing more harm than good. In work published from 2005 onwards, Karikó and Weissman figured out how to create a hybrid mRNA which wouldn't trigger that immune response. To quote a 2021 article by Nicoletta Lanese for the science news website *Live Science*:

> In essence, Karikó and Weissman figured out how to quiet alarms from the immune system long enough for synthetic messenger RNA to slip into cells, send commands to the cells to make proteins, and be broken down harmlessly once those instructions were delivered. That process enabled the COVID-19 vaccines [...] and the technology could pave the way for gene therapies and cancer treatments, in the future.

By now Karikó and Weissman's work has certainly received the recognition it deserves. At the time of writing the first draft of this book they had won four of the biggest prizes going in life sciences – the Horwitz, Albany, Lasker and Breakthrough Prizes, and added the Nobel Prize for Physiology or Medicine in October 2023.

Recombinant DNA

This is cutting-edge stuff but mRNA is just one of many similarly exciting technologies to have emerged in the last several decades as molecular biology has evolved. Another is that of 'recombinant DNA'.

Earlier, we mentioned in passing that the first large molecule/'biologic' drug ever approved was synthetic human insulin sold under the brand name 'Humulin' by the large pharmaceutical company Eli Lilly. Humulin was launched in 1982. Prior to this development of synthetic insulin, diabetics all over the world had been treated with insulin obtained from animals. Frederick Banting and John Macleod, both of Toronto University, were awarded the Nobel Prize in Physiology or Medicine in 1923 'for the discovery of insulin'. Strictly they didn't really discover insulin, but they were the first to successfully extract it and use it to treat a diabetic. From that point onwards, insulin was taken from cattle and pigs. Eli Lilly produced enough insulin for the US market from the 1920s onwards. On the other side of the Atlantic, two Danish companies, Nordisk Insulinlaboratorium and Novo Terapeutisk Laboratorium, took the work done in Canada and began to produce insulin for the European market.

This is yet another example of the pharmaceutical industry delivering what might be described as a 'miracle cure'. For hundreds of millions of patients from then until the present day, diabetes was no longer a death sentence.

Over the next many decades Eli Lilly, the two Danish firms and the German company Hoechst (now part of the French company Sanofi) gradually refined and improved the product, though it took nearly 60 years before the big leap forward which was the manufacture of *synthetic human* insulin. That step came in large part as a result of the work done on DNA. In the 20 years or so following Watson and Crick's Nobel Prize, scientists gradually worked out how to combine DNA from at least two sources, to then engineer a desired protein. This process became known as 'genetic recombination'.

By 1982 scientists at the American company Genentech discovered how to insert the human insulin gene into the bacterium *Escherichia coli* (*E. coli*) which enabled them to then manufacture human insulin. Given their expertise in the area, Eli Lilly licensed this intellectual property and began to manufacture synthetic human insulin at scale in a factory rather than have to extract it from an animal source. Pigs and cows everywhere no doubt breathed a huge sigh of relief (and will breathe an even bigger one should the vision of perfect, 'synthetic' cultured meat be delivered in the relatively near future). Even better, the supply of insulin was now essentially unlimited.

Recombinant DNA is another technology the development of which rested on progress made over several Nobel Prizes. Today it is the foundational science which underpins a number of therapeutic treatments and diagnostic technologies.

Monoclonal antibodies

Yet another Nobel Prize-winning technology to have emerged in the last few decades on the back of our increased understanding of and ability to manipulate and clone DNA is that of 'monoclonal antibodies', or 'mAbs' for short. These are the most common kind of 'large molecule' or 'biologic' drugs described earlier.

In March 2021 the US FDA approved its hundredth monoclonal antibody product. Together, those products generated more than $180 billion of revenues in 2021, and this number is forecast to grow to some way more than $400 billion a year by the end of the decade. This is perhaps unsurprising when we consider that by now such products have demonstrated efficacy in treating transplant rejection, preventing blood clots and against a wide range of difficult diseases, including rheumatoid arthritis, asthma, Crohn's disease and ulcerative colitis, multiple sclerosis, macular degeneration and a great deal else. Perhaps of most importance, however, they have 'revolutionized cancer treatment and [...] fundamentally changed the way we view how cancer can be managed'.

I don't think it is particularly controversial to suggest that most people were broadly familiar with the idea of antibodies even before the COVID pandemic brought them even further into the public consciousness. They are Y-shaped proteins that form part of the front line of our body's immune system, along with 'B-cells', 'T-cells' (together 'lymphocytes'), and 'macrophages' and 'neutrophils' (both 'phagocytes' – cells which engulf pathogens to remove them from the body). Each of these performs a specific function to deal with foreign 'invaders' or 'pathogens' in the body such as bacteria, viruses or fungi. The job of an antibody is to lock on to an 'antigen', which is a molecule or molecular fragment on an invading pathogen which the antibody is able to identify as 'alien' so that the rest of the immune system can then get rid of that pathogen.

As scientists began to understand the function of antibodies and antigens, it seemed entirely logical that the ability to produce antibodies with predetermined specificity could lead to powerful therapeutics. If we could identify an antigen carried by a given pathogen and then manufacture the corresponding antibody, we could use that as a highly effective treatment for that pathogen. This was clearly an extraordinarily complex challenge. Not only would scientists need to be able to identify antigens with exquisite precision, but they would also need to understand how to produce the corresponding antibody. Perhaps just as important, for this to work as a treatment, they would also need a way of manufacturing that antibody at scale and at a cost that wasn't too much for healthcare systems to bear.

Happily, from the 1970s onwards that is precisely what happened. In 1975, based at Cambridge University's Laboratory of Molecular Biology in the UK, Georges Köhler (1946–95) and César Milstein (1927–2002) worked out a way to produce targeted 'monoclonal' antibodies which could recognize one specific antigen. Their approach was called the 'hybridoma' technique. Their key insight was to fuse antibody-producing white blood cells with tumour cells (to produce 'hybrid' cells, hence the name). White blood cells can't survive outside of the body for very long, which meant they couldn't be manipulated or used to manufacture antibodies.

Fusing those cells with myeloma cells resulted in an 'immortalized' hybrid cell which could survive and, crucially, also continue dividing, thus providing an unlimited source of specific antibodies.

Köhler and Milstein received the 1984 Nobel Prize for their work. The Nobel Institute's press release described their new technique as 'the most important methodological advance within the field of biomedicine during the 1970s'.

The hybridoma technique necessarily required the use of animals, most often mice but also rabbits and the proverbial laboratory 'Guinea pig' too. Although the technique was a significant breakthrough, the human immune system was always going to have a problem with antibodies derived from a non-human source. It took another few years for the next development which was needed before monoclonal antibodies could really take off for use in human therapeutics. Working in the very same University of Cambridge laboratory as Köhler and Milstein, from the early 1980s onwards, Gregory (now Sir Gregory) Winter (1951–) and his team set about working out how to 'humanize' monoclonal antibodies. By the end of the 1980s they had achieved just that. Using new techniques, they were able to produce antibodies that were genetically human, thereby reducing or eliminating problems associated with the early non-humanized antibodies.

One such technique was called 'phage display', which had originally been developed from 1985 onwards by George Smith (1941–) at the University of Missouri. Winter and his team developed phage display further in Cambridge and used it in their work on humanized antibodies. Winter and Smith were two of the three scientists to share the 2018 Nobel Prize for Chemistry for that work as a result. As the 2018 Nobel press release put it:

> Winter used phage display for the directed evolution of antibodies, with the aim of producing new pharmaceuticals. The first one based on this method, adalimumab, was approved in 2002 and is used for rheumatoid arthritis, psoriasis and inflammatory bowel diseases. Since then, phage display has produced anti-bodies that can neutralise toxins, counteract autoimmune diseases and cure metastatic cancer.

Adalimumab's brand name is Humira, sold by the American pharmaceutical company AbbVie. In 2022 AbbVie sold more than $21 billion worth of the drug globally. This is the economic value that Nobel Prize-winning science can and does create over time.

Immune checkpoint inhibitors (ICIs)

Interestingly, in the very same year that Sir Gregory Winter and George Smith were awarded the Nobel Prize in Chemistry, the Nobel Prize in Physiology or Medicine was awarded for yet another key development in the world of monoclonal antibodies, the discovery of which rested on the work already done by Köhler, Milstein, Winter, Smith and many others. This was the science which, to some degree at least, was able to 'cure metastatic cancer' as per the press release for the Chemistry prize quoted above. In 2018 James P. Allison (1948–) and Tasuku Honjo (1942–) won their Nobel Prize 'for their discovery of cancer therapy by inhibition of negative immune regulation'. Specifically, their work enabled an entirely new approach to cancer treatment called 'immune checkpoint therapy'.

For years, scientists have understood that our immune system has a delicate balance of 'brakes' and 'accelerators' when it comes to dealing with foreign invaders. This ensures that our immune system will only attack something 'bad' when it should (by 'accelerating' our various cellular soldiers, such as T-cells) and will leave 'good' cells from our own tissues and organs alone (by applying certain 'brakes'). As the Nobel Institute has put it: 'This intricate balance between accelerators and brakes is essential for tight control. It ensures that the immune system is sufficiently engaged in attack against foreign microorganisms while avoiding the excessive activation that can lead to autoimmune destruction of healthy cells and tissues.'

One of the biggest impediments to cancer treatment has been the fact that cancer cells are very good at 'hiding' from our immune systems which often struggle to recognize them as 'alien'. Working thousands of miles away from one another, with Allison in Berkeley, California, and Honjo in Kyoto, Japan, both scientists were focused on the idea that this happens because certain proteins act as brakes on our immune systems, particularly when it comes to cancer. Specifically, Allison was focused on a protein called 'cytotoxic T-lymphocyte-associated protein 4', or 'CTLA-4' for short, and Honjo on the rather dramatically named 'programmed cell death protein 1', or simply 'PD-1'.

The scientists realized that both of these proteins were applying brakes to our immune systems which meant that cancer cells would all too often evade destruction and proliferate. They theorized that, if you could block or inhibit those protein brakes, you could potentially 'unleash the immune system to attack cancer cells'. To do this 'all' you had to do was develop an antibody that could bind to CTLA-4 or PD-1.

Allison developed just such an antibody for CTLA-4 in the early 1990s. In 1994 he and his team first used it in experiments with laboratory mice. The results were spectacular: Allison and his team were able to effectively cure cancer in mice that were treated with the CTLA-4 antibody.

This was a first step on a long road between then and now. As hard-bitten biotech chief executives, clinicians and professional investors will tell you, it is pretty well established by now that it can be surprisingly easy to cure cancer in mice. That outcome is achieved fairly often and can then be crowed about by mainstream journalists who get excited rather prematurely. It is very much harder to cure cancer in human beings. By 2010, however: 'an important clinical study showed striking effects in patients with advanced melanoma, a type of skin cancer. In several patients signs of remaining cancer disappeared. Such remarkable results had never been seen before in this patient group.'

Similarly exciting results were achieved in PD-1 trials. Roll on to the present day and pharmaceutical company Merck's Keytruda, the top-selling PD-1 antibody drug in the world, generated nearly $21 billion of revenues in 2022. The drug is approved for treatment of patients with several cancers, including melanoma, non-small cell lung cancer (NSCLC), head and neck squamous cell cancer (HNSCC), Hodgkin's lymphoma, bladder cancer, colon and rectal cancer, and several others, hence the reason it is such a top-selling drug treatment.

This broad approach is called 'immunotherapy' because it uses the body's own immune system to fight cancer. Cancer immunotherapy has been one of the most exciting and valuable developments in biotech of the last few years. Not only can it be highly effective but, when used as a monotherapy (i.e., on its own), it often has less severe side effects than those we all know are too often suffered with traditional therapies such as disfiguring surgery, radiation or chemotherapy.

That said, there are still real challenges with these ICI drugs. First, response rates vary a great deal across different patients and different cancers. For some lucky patients, a 'complete response' (CR) is achieved. That is to say that to all intents and purposes their cancer is 'cured'. For many, however, there is only a 'partial response', meaning that their cancer might stall or progress more slowly but is by no means cured. This can still be a positive outcome for such patients as they can then enjoy several more months or even years of 'progression free survival' (PFS) – a great outcome for them and for their loved ones. For others, the treatments may not work at all. These drugs certainly constitute a significant forward step in cancer care but they are as yet far from a panacea, and it is certainly not appropriate to speak of them as an effective 'cure for cancer'.

Another key problem with ICI drugs is that they are expensive, often very expensive. Keytruda's list price in the USA is above $100,000 a year, although healthcare systems and insurers all over the world pay different prices which they negotiate directly with Merck and which are based on a complicated cocktail of considerations in terms of what the local market can bear price wise and the ability of centralized systems to secure lower prices because of the volumes they can buy. In June 2018, for example, the National Health Service (NHS) in the UK announced that they had concluded a deal with Merck to make Keytruda available to British patients. The price agreed was confidential but certainly some way lower than the list price in the USA.

For some, this is an example of the merits of the British system but it is probably worth highlighting that not only was this agreement reached fully two years after Keytruda had been approved in the USA but even to this day the drug is only available to a small number of patients in the UK and in a limited number of indications – that is, to treat a limited subset of the cancers for which Keytruda is an approved treatment elsewhere.

Other ICI drugs are similarly expensive. US pharmaceutical giant Bristol-Myers Squibb (BMS) was one of the early entrants into the market with a drug called Yervoy, which was a CTLA-4 antibody approved to treat metastatic melanoma as long ago as 2011. BMS then followed up with its own PD-1 drug, Opdivo, in 2014, the same year that Merck's Keytruda was approved. Swiss giant Roche followed up with its PD-L1 antibody, Tencentriq, in 2016.

All of these drugs cost healthcare systems and insurers comfortably north of six figures. The real picture is actually even worse in terms of cost because most of them have tended to be used in *combination* with other treatments rather than on their own as a 'monotherapy'. Because of their limited efficacy, in the time since these drugs have come on to the market, clinicians have tended to use them in combination with existing standards of care such as chemotherapy, surgery and radiotherapy or even, in some cases, in combination with each other.

As an example, Yervoy and Opdivo have been approved for use together in combination for patients with advanced or inoperable melanoma for some years. The idea was that using CTLA-4 and PD-1 antibodies together could be even more powerful than using one or other of them on its own. A combination of Yervoy and Opdivo would do just that. The resultant combination therapy would run to more than $250,000 per annum, however. As a 2017 Reuters article reports:

> 'For cancer drugs in general [...] it is hard for us to drive down cost,' said Steve Miller, chief medical officer at Express Scripts Holding Co, the nation's largest manager of drug benefit plans for employers and insurers. 'You don't want to be in the position of being told to use the second best cancer drug for your child.'

These sentiments are perhaps unsurprising when you consider the clinical results of that combination delivered by BMS in 2021. Over a clinical trial lasting six and a half years, 49 per cent of the patients treated with Opdivo plus Yervoy were still alive and 77 per cent of those patients 'remained treatment-free'. From a potential death sentence in around six months to surviving for several years and needing no further treatment for a meaningful percentage of patients – that is certainly reasonable progress. For the lucky few who have the insurance coverage to be treated at a cost of more than $250,000 and who respond to the treatment, this kind of outcome is obviously rather wonderful.

More exciting, however, is what the future may hold. We are only just about one decade into the development and use of these drugs. There is a good chance that the science will continue to improve and, importantly, that manufacturing costs will come down enough to make these sorts of treatments significantly cheaper and far more widely available in the years ahead as a result.

The reason these drugs are so expensive is that they are incredibly complicated and costly to manufacture. Monoclonal antibodies, including these ICIs, are large molecule drugs. As we saw earlier, these are significantly more complicated to build and manufacture than their small molecule counterparts. They are also less stable which means that they require more complicated supply chains. When you consider the approaches required to make even small quantities of these drugs, the processes, tools and equipment and highly specialized human resource needed, for example, it becomes relatively easy to understand why they tend to be so eye-wateringly expensive.

Bi-specific/multi-specific antibodies

In fact, one of the more recent trends in the area of immunotherapy is an attempt to make even more complicated molecules. Many of the leading antibody drugs of today are what are called 'monomers'. They are designed to target *one* single antigen such as CTLA-4 or PD-1, as we have seen. For some time now, clinicians have experimented with using more than one 'monomeric' antibody together, as in the example above with Yervoy and Opdivo.

All other things being equal, it was assumed that targeting two antigens could be more powerful than targeting just one. This assumption was then confirmed to a great extent by 'combination' trials such as the Yervoy and Opdivo one referenced above. Given this outcome, as the science advanced, another logical idea which arose was that of building 'bi-specific' or even 'multi-specific' antibodies. That is to say antibody molecules which could bind to more than one antigen themselves so you would only need one drug to deal with two (bi-specific) or even more targets (multi-specific). As the US biotech company Amgen explains:

> Bispecific antibodies aim to treat multifaceted, complex diseases by engaging two disease targets with one molecule. While natural antibodies have two targeting arms that bind to the same target antigen, bispecific antibodies are engineered hybrid molecules with two distinct binding domains that target two distinct antigens.

Amgen's drug 'BLINCYTO' (blinatumomab) was the first 'bispecific' drug to be approved by the FDA as long ago as 2014 for the treatment of patients with an aggressive form of leukaemia who had not responded to other treatments. BLINCYTO is able to bind two distinct molecules, CD19, which is found on the surface of leukaemia cells, and CD3, which is found on our own T-cells. As Dr Elad Sharon of the US National Cancer Institute explains: 'Essentially, the drug gets the immune system to notice – and then target – the cancer cells.'

In follow-up trials conducted since 2014, the drug has shown efficacy and has been proven to be much less toxic than standard chemotherapies. But this outcome comes at a price, of course. When it was launched, BLINCYTO was one of the most expensive cancer drugs on the market, with a list price of $178,000.

In summary

In this chapter we have examined how the biotech industry has developed with specific reference to therapeutics, that is, to drugs which treat disease. We have come a long way in the last few centuries from hit-or-miss treatments based on natural extracts, to and through the development of small molecules, vaccination, penicillin and anti-infectives and on to the more explicitly 'biotech' technologies of recent decades: DNA, RNA, mRNA and the large-molecule drugs which have resulted.

In the next chapter we will look at a number of other cutting-edge technologies including gene and cell therapy, stem cells and gene editing, and at why eye-watering prices for all of these kinds of drug treatments can actually make sense for healthcare systems and for society as a whole.

9
Health economics and the latest treatments

We ended the last chapter with a focus on some very expensive drug treatments. Before we look at some even more expensive areas of biotech innovation at the very cutting edge of science, it is probably worth a brief digression to explain how many of the world's healthcare systems tend to think about health economics and drug pricing and, specifically, their use of something called a 'QALY'.

Health economics and QALYs

A QALY is a 'quality-adjusted life year'. Health economists and clinicians seek to ascribe a score from 0 to 1 where 0 = death and 1 = one year in decent health. If, after a given treatment, someone lives an entire year in good health, the treatment would have achieved a score of '1'. If someone is able to survive six months after a treatment in good health and then dies, the score ascribed is 0.5. Similarly, were they to live a year but with poor health that was more or less subjectively evaluated as 'half' as good as 'good' health, that would also deliver a score of 0.5. Two years in perfect health would be scored at 2, three years at 'half' of good health at 1.5 and so on.

Different healthcare systems ascribe different monetary values to a QALY. In the USA, the Institute for Clinical and Economic Review tends to value one QALY at anywhere between $50,000 and $150,000. Elsewhere in the world, the numbers tend to be somewhat lower. In the UK, in the past the National Institute for Health and Care Excellence (NICE) tended to use a threshold of somewhere between £20,000 and £30,000, although this has been increasing and nowadays it will give consideration to more expensive treatments on a case-by-case basis, as it has with cancer drugs in recent years.

As cold and calculating as the idea of a QALY may sound, it is a framework which can be used in an attempt to quantify the cost-effectiveness of a new medicine as against an existing standard of care, and, to be fair to health economists, medical systems, insurers and government ministers everywhere, we do need some way of making substantive decisions about how to deploy limited resources when giving consideration to medical treatments.

As an example of how this can work, a 2017 *Journal of Medical Economics* article by Thomas Delea and colleagues investigated the cost-effectiveness of the aforementioned BLINCYTO versus the existing 'salvage' chemotherapy standard of care. That study found that using the expensive drug was likely to be cost-effective even given the high price of the drug. Valuing a QALY at $150,000, the drug delivered some way more in terms of the value implied by QALYs as against using chemotherapy.

The advantage in that specific example was perhaps somewhat marginal, but it isn't hard to find far more compelling cases where 'expensive' drugs can deliver significantly improved health economics overall, and this is important to note when we consider drug pricing.

In any debate around drug pricing generally, it is also important to consider the fact that total drug spending averages only about 15 per cent of healthcare expenditure overall. A visualization for this is available at https://www.iqvia.com/insights/the-iqvia-institute/reports-and-publications/reports/drug-expenditure-dynamics

Crucially, there is a fair bit of evidence that spending on drugs can often drive down the other 85 per cent of healthcare costs. As Jean-François Formela and John Stanford put it in an April 2022 article in *STAT News*: 'Prescription drugs can significantly reduce the need for expensive emergency room visits, surgeries, hospitalizations, and long-term care.' In that same article they go on to provide a particularly powerful example of this phenomenon at play. As they explain:

> Barely a decade ago, 20 per cent of people with hepatitis C would develop cirrhosis, a complex and expensive condition that can necessitate a liver transplant. Today, there are once-daily medications that can cure up to 95 per cent of cases with few to no side effects. A course of one of those medicines costs $24,000. That is certainly not cheap, but it is one-twenty-fifth the cost of a liver transplant, which costs $600,000 on average.

It is these sorts of health economic and quality of life considerations that have enabled an even more expensive class of drug to emerge in the last few years: cell and gene therapies.

Cell therapy and gene therapy

Two other exciting and relatively new technologies that ultimately rest on much of the same foundational research which produced large molecule drugs such as antibodies and checkpoint inhibitors are those of gene therapy and cell therapy.

The US FDA defines gene therapy by explaining that it 'seeks to modify or manipulate the expression of a gene or to alter the biological properties of living cells for therapeutic use'. It goes on to explain three broad mechanisms of action, specifically: 'replacing a disease-causing gene with a healthy copy of the gene', 'inactivating a disease-causing gene that is not functioning properly' and 'introducing a new or modified gene into the body to help treat a disease'. As we have come increasingly to understand genetics in the last couple of decades, this broad approach has finally become a reality, and there are already a small number of treatments being used with patients in the real world.

Cell therapy is closely related to gene therapy but involves the use of intact, live cells to treat a disease. Such cells are usually engineered outside of the body ('ex vivo') and will either have come from the patient originally and have been re-engineered (known as 'autologous' cells), or from a donor of some kind ('allogeneic' cells).

Slightly confusingly, therapies in this broad area can often be considered both gene and cell therapies at the same time. This is less complicated than it sounds as it simply means that the treatment works by altering or 're-engineering' the *genes* in the *cells* which are then inserted back into the patient as the treatment. This is sometimes described as 'cell-based gene therapy'.

Arguably, one of the highest profile of these types of therapies to date is called chimeric antigen receptor T-cell therapy, or 'CAR-T' for short. CAR-T works by removing white blood cells from a patient via a process called 'leukapheresis'. T-cells are then separated out and engineered in a laboratory to add a gene which will help those T-cells identify and attack a specific cancer. These cells are then expanded over several days so that there are many millions of them before they are then infused back into the patient.

This is philosophically quite similar to the approach used with PD-1 and CTLA-4 antibodies we looked at a few pages back. In certain kinds of cancers, specifically leukaemia and lymphoma, the cancer cells carry an antigen called 'CD19'. CAR-T cells are engineered to bind to the CD-19 antigen just as PD-1 antibodies are engineered to bind to PD-1.

The first person in the world to undergo this treatment just before her seventh birthday in April 2012 was a little American girl named Emily Whitehead. Emily had been diagnosed with acute lymphoblastic leukaemia (ALL) in May 2010 and had not responded to conventional chemotherapy after several rounds of treatment. By early 2012 her parents were told she probably had only a few weeks to live. They decided to enrol her in a Phase I CAR-T clinical trial being conducted at the Children's Hospital of Philadelphia (CHOP) working in collaboration with Swiss pharmaceutical giant Novartis.

The treatment worked, and on 10 May 2012 Emily Whitehead's life-threatening, late-stage cancer went into remission. In 2022 Emily celebrated ten years 'cancer free'. In the time since her treatment Emily and her parents have set up the Emily Whitehead Foundation to help raise funding for further research into cutting-edge cancer treatments and to share their message given how relatively few people are as yet aware of CAR-T therapy.

It took another few years after the success of Emily's treatment for Novartis to take the approach all the way through clinical trials and gain regulatory approval for their CAR-T drug Kymriah, which was approved by the US FDA in the summer of 2017. In June 2022 Novartis published its five-year data for Kymriah used in children and young adults for the treatment of acute lymphoblastic leukaemia (ALL). Fifty-five per cent of the patients were still alive after more than five years and 44 per cent were still in remission. As Stephan Grupp, the doctor who led the trial at the Children's Hospital of Philadelphia (CHOP) put it:

> These data mark a moment of profound hope for children, young adults and their families with [...] ALL, as relapse after five years is rare. [...] Since the approval of Kymriah nearly five years ago, we have been able to offer a truly game-changing option to patients who previously faced a five-year survival rate of less than 10 percent.

Since Kymriah's approval in 2017, a number of other CAR-T therapies have been approved. Not only have they delivered 'fantastic' results in children with leukaemia, but they have also been approved to treat certain patients with aggressive lymphomas, another kind of blood cancer. In the few years since they have been approved, CAR-T therapies have arguably rightly been described as a revolution in cancer care. One study published in February 2022 even found that two of the first adult patients to be treated in early clinical trials as long ago as 2010 remain cancer free and, crucially, still have cancer-fighting CAR-T cells in their blood today more than ten years after their initial treatment.

This is all very encouraging but, as ever, there are caveats. First, while some patients are lucky enough to enjoy complete remission and many years of progression-free survival as we have seen, there are still a meaningful number of patients who do not. Perhaps more important, however, is that CAR-T therapies cost even more than the immune checkpoint inhibitor drugs we covered earlier. When Novartis first brought Kymriah to the market, the price was set at $475,000 for a single infusion. Gilead's drugs Yescarta and Tecartus both came to market at a list price of $373,000, and two CAR-T treatments for multiple myeloma approved in 2021 and 2022, BMS's Abecma and J&J's Carvykti, cost $441,000 and $465,000, respectively.

Unsurprisingly, this has meant that such treatments have only ever been available to a relatively small number of patients in the developed world and basically only for patients whose cancer is 'relapsed or refractory', that is to say patients where more conventional treatments have been tried initially but whose cancer keeps coming back or has stopped responding to treatment. In the vernacular of cancer treatment, CAR-T therapies are 'last-line' treatments which are generally only used once several other more conventional treatments have been attempted and have failed.

While these treatments remain so expensive, there is limited scope for them to be used far more widely and as 'first-line' therapies instead of existing standards of care such as the chemo- and radiotherapy which have such awful side effects. Success here might mean that it won't be too long before patients could be spared two years or more of painful and unpleasant chemotherapy treatment.

It is entirely possible that exponential progress could continue to hone and improve these sorts of treatments, thus increasing their efficacy and, hopefully, bringing costs and prices down to a level which would enable their use to be far more widespread. Assuming this trajectory, such drugs could even get to the point where they may eventually replace those existing standards of care which have such horrendous side effects for so many patients.

It may take a few years yet, but this outcome should be possible given the march of scientific progress. As an example of why I believe this – when it comes to CAR-T specifically – their production requires something called a 'viral vector' to introduce the genes into the T-cells which can then turn them into CAR-T cells to fight the cancer. The British company Oxford Biomedica – my former client – is a leading player in this area and was the supplier of a particular kind of viral vector, called 'lentiviral vector', which was crucial to the development and manufacture of Kymriah. As it worked with Novartis over several years from 2013 onwards to be ready to produce the drug for patients, it was able to improve its processes such that it could bring the manufacturing costs down tenfold and achieve a tenfold increase in production yields as compared to its initial delivery technology. Oxford Biomedica shared this outcome publicly in its interim results announcement in June 2017.

This brought the 'cost of goods' for the drug down from a level which would have made it enormously loss-making to a level which made it possible to make the drug available to patients even if the price was still as high as $475,000 to begin with. At the time Kymriah was launched, many leading healthcare analysts were still unsure as to whether the drug would be profitable for Novartis, even at that price, particularly once you factored in the need to recover the very significant sums spent on research and development and the many years of work required to invent the drug and get it all the way through to approval.

Since then, however, Oxford Biomedica has stated publicly that, thanks to continued innovation in its manufacturing process, it can continue to reduce costs and improve yields, particularly as it has scaled its production facilities and headcount to deliver contracts with several more companies working in the area. In March 2019, it also announced a research and development collaboration with Microsoft: 'to improve the yield and quality of next-generation gene therapy vectors using the cloud and machine learning'.

Manufacturing costs are confidential for commercial reasons, given what a highly competitive industry this is, but based on the extent to which the company has been able to reduce costs and increase yields, it seems likely that in the last decade or more Oxford Biomedica has probably managed to bring the manufacturing cost per treatment down from a high six- or even low-seven figure sum to tens of thousands of dollars per treatment.

This phenomenal achievement by Oxford Biomedica and other companies like it suggests that we may be on a glide path towards these sorts of cutting-edge treatments being far more affordable and widely available in the fullness of time. Such progress also brings on to the horizon the possibility of being able to treat a much wider range of diseases than we can at present. Many cancers require significant quantities of 'lentivector' to deliver a sufficient therapeutic 'payload' to fight a given disease. With six- or seven-figure unit costs for lentivector there was no chance we would ever be able to address liver or lung cancer for example. Too much viral vector would be required, and the costs would therefore be entirely prohibitive.

As companies like Oxford Biomedica and Microsoft have continued to work on such things, however, the likelihood such technologies will be applicable to far more diseases in gene and cell therapy increases. On that point, it is worth quoting Andrew Phillips, head of Biological Computation at Microsoft, in full. As he put it in a March 2019 press release:

> Programming biology has the potential to solve some of the world's toughest problems in medicine, and to lay the foundations for a future bioeconomy based on sustainable technology. Oxford Biomedica is at the cutting edge of cell and gene therapy delivery and their highly sophisticated manufacturing processes generate a vast wealth of valuable data. We anticipate that by combining computational modelling, lab automation, machine learning and the power of the cloud, we can help them in their quest to make existing treatments more cost effective and in future to develop ground-breaking new treatments.

Even at $475,000 the health economics of such treatments can actually make sense. First, when you consider things in terms of QALYs, particularly in paediatric patients. With one treatment you could potentially add decades of healthy lifespan to a young child like Emily Whitehead suffering from leukaemia. If a QALY is valued at $150,000, for example, then the economics are compelling given that so many patients are experiencing progression-free survival five or even ten years after treatment. Parents of a child suffering from that sort of condition would certainly agree!

The other reason the health economics of such therapies can stack up is that existing standards of care can be extremely expensive anyway. A bone marrow transplant can cost as much as $900,000, for example, and has the additional complication of the need to find a matching and willing donor.

How exciting, however, to contemplate a world where these sorts of treatments become very significantly less expensive over time and far more widely available as a result. If the lesson of history and the direction of travel here are anything to go by, we may be able to look forward to precisely this outcome. It may not be too many years before big pharmaceutical companies will be able to sell 'miracle cures' at prices which will make them widely available for many more diseases and even, dare we hope, make them affordable in the developing world eventually.

In the meantime, however, such companies really do need to make enough money to fund the many billions they spend on research each year to develop these treatments still further, on staff and on manufacturing facilities and processes which can cost hundreds of millions. Rather than criticize biopharma companies for these sorts of eye-watering drug prices, we should perhaps think a bit more deeply about the merits of those treatments, even at such prices, and realize that this is how capitalism works and how we fund innovation and giant leaps forward scientifically at the most fundamental level. It is this very dynamic which has given us the incredible progress we have enjoyed in the last two centuries in so many key areas of human development. If we see these organizations as evil, rapacious 'profiteers', we risk strangling innovation and very significantly slowing the progress we could be making or even stopping it altogether.

Air travel was incredibly expensive when it first emerged and only available to a lucky and very wealthy few. Now it is enjoyed by billions. The same was true of most of the other consumer products and services which so many of the world's population can and do enjoy today: televisions, computers, cars, smartphones and so much else besides. The processing power in today's smartphone would have cost millions only a generation or so ago, yet today we have got the cost down far enough for several billion of them to have been sold throughout the world.

Today's most innovative drugs are expensive because they are at the very cutting edge of our best technology. This costs a great deal, just as the first planes, televisions, computers and cars did decades ago. It is also crucial to note that such progress needs to be profitable to reward shareholders for taking the risk of investing in the companies working on such things else there wouldn't be the capital available to do any of it and no progress could be made.

I have first-hand experience of watching extraordinarily dedicated biotech management teams, research staff and clinicians work tirelessly and in the face of endless challenges and complications to bring costs down to a level which could make these kinds of treatments available to patients, and then carry on with that progress thereafter in a bid to make them more widely available.

Allogeneic cell therapy

Another reason that prices may come down in time is because any number of other technologies could end up being fundamentally cheaper than the approaches taken today. One of the reasons CAR-T therapy is so expensive at present, for example, has to do with the fact that the patient's own cells are used – the approach uses 'autologous' (patient-derived) cells.

This introduces a great deal of cost and complexity into the process. A given patient will have to be able to travel to a medical facility which is capable of performing leukapheresis (the removal of white blood cells from the patient) and which may be some distance from the patient's home. There are then significant shipping and manufacturing times required before the 'living drug' is ready and can be returned to the patient for infusion. This can take several weeks, and the entire process can therefore add a great deal of cost, complexity and risk, especially given how sick many of the patients are.

An alternative approach is to engineer 'off the shelf' CAR-T cells from healthy donors ('allogeneic' cells). These would need no such complex supply chain, could be manufactured in bulk rather than for each individual patient, and could also be given to a patient the moment they arrive at a treatment centre. This could significantly improve the patient's experience and decrease the cost per patient by some margin.

Allogeneic treatments have certain inherent challenges, most particularly immune rejection by the patient. At best, rejection of an allogeneic cell–based treatment would limit or even eliminate the efficacy of that treatment. At worst, it could make the patient extremely sick and potentially induce something called 'graft versus host disease', or 'GvHD', a life-threatening condition where the donated immune cells could attack the patient's body.

There is a great deal of research being conducted in this area, however. Via a process known as 'gene editing', scientists are working to genetically engineer allogeneic cells such that they won't trigger immune rejection. As a 2021 article by Kenneth J. Caldwell and colleagues in *Frontiers in Immunology* puts it:

> Gene editing has emerged as the leading strategy being tested in the clinic as investigators seek to develop an 'off the shelf' allogeneic CAR therapy platform. As gene editing techniques advance, this method offers great potential to mitigate potential risks and downsides associated with the use of allogeneic cells.

In vivo

These autologous and allogeneic approaches are both 'ex vivo' approaches, in that the therapy is engineered outside of the body. Another exciting idea is to trigger the production of CAR-T cells in the body itself – 'in vivo'. The science is very early stage but there are already groups working on the delivery of modified messenger RNA (mRNA) into the body to reprogram T-cells. This 'mRNA-based, transient CAR-T cell technology' is another approach which could significantly reduce costs and be used to address many diseases in the fullness of time.

Stem cells

We can't move on from the subject of cell and gene therapy without covering stem cells specifically. Many more people will likely have heard of stem cells and stem cell research than have heard of CAR-T therapy or monoclonal antibodies for the simple reason that the area has attracted significant controversy in recent decades.

Stem cells are one of the key building blocks of multicellular organisms. Although this is a gross oversimplification, the vast majority of cells in our bodies are made to do one specific job. They will be specialized to make up our blood, liver, cartilage, muscle, bone, skin and so on. In contrast, stem cells have two broad properties. First, they have the capacity to 'differentiate' into any number of those more specialized cells, something referred to as 'potency'. Second, they are capable of division and proliferation such that they can create copies of themselves as well as generating those specialized cells. Dividing stem cells create 'mother' cells which are copies of the original cell and which have the same reproductive capacity and potency and 'daughter' cells which are those differentiated cells produced to perform a specific function within the body.

When a sperm fertilizes an egg in reproduction, the resulting zygote – the very first cell of an individual's existence – is a 'totipotent' or 'omnipotent' stem cell. These cells have such 'potency' that they are capable of dividing to create an entire functional living organism – including you and me! These totipotent cells then develop to create a series of different types of stem cell in descending order of potency. Roughly four days following fertilization, totipotent stem cells begin to specialize into 'pluripotent' stem cells. These don't have quite the same 'creative power' as the totipotent cells that preceded them but they are nevertheless capable of self-renewing and differentiating into any of the three main categories of cells required to build essentially all of our tissues and organs.

Three further primary categories of stem cell which follow thereafter are 'multipotent' cells which can differentiate into a reasonably wide range of different underlying cell types – 'oligopotent' cells which can differentiate into a more limited range of 'daughter' cells and 'unipotent' cells which can also self-renew but can only 'manufacture' a single cell type rather than a wider range of different ones.

These stem cells are crucial for the production and renewal of all of our cells throughout the body during the course of our lives. The five different categories of stem cell are important to understand since each of them will have different potential therapeutic application given their own unique properties.

Mesenchymal stem cells (MSCs), for example, are a kind of multipotent cell which can differentiate into bone, muscle, fat and cartilage. Hematopoietic stem cells (HSCs) are an oligopotent kind of stem cell which can differentiate themselves into various blood cells, including the white blood cells so critical for our immune system such as T-cells.

As may seem more or less obvious given what these cells do, ever since their discovery there has been great excitement from research scientists and clinicians alike about their potential use. The observation of stem cells can help us better understand how diseases develop and stem cells can also be used to test new drugs before being used in patients. Most exciting of all, however, is the developing field of 'regenerative medicine' where 'fresh' stem cells could be used to regenerate and repair organs and tissues which have been damaged by disease or trauma.

Although it is still relatively early days in the field, in terms of approved treatments at least, the area has experienced explosive growth in recent years. There are currently several thousand clinical trials investigating the use of stem cells in an astonishingly wide range of diseases and health conditions, including but not limited to: Alzheimer's, arthritis, bone regrowth, burns, cerebral palsy, Crohn's disease and ulcerative colitis, cystic fibrosis, diabetes, various lung and liver diseases, hearing loss, multiple sclerosis, spinal cord injuries and stroke. The list is very long indeed, and this is one broad area of research and scientific progress which could well deliver some astonishing results in the relatively near future given the pace of progress and the sheer volume of research and number of trials being conducted all over the world.

Hematopoietic stem cell transplantation

A treatment called 'hematopoietic stem cell transplantation' (HSCT) has actually already been in use since the mid-1950s. This procedure was initially used to treat certain blood cancers and cancer of the bone marrow and used HSCs from the patient's own bone marrow (autologous) or from the bone marrow of a healthy, willing and matching donor (allogeneic). In 1990 E. Donnall Thomas (1920–2012) and Joseph Murray (1919–2012) shared a Nobel Prize in Physiology or Medicine for their work in this area.

As our understanding of stem cells progressed, however, it was realized that umbilical cord blood was another rich source of HSCs. Securing HSCs from umbilical cord blood is rather easier and less painful, dangerous and invasive than from bone marrow. Harvesting stem cells from bone marrow involves a large needle being inserted into the back of a patient or donor's hip bone to pull out the viscous liquid marrow. This is a reasonably serious procedure involving either a general anaesthetic or spinal anaesthesia and can result in significant 'post-harvest' pain and fatigue. By contrast, umbilical cord blood can be collected reasonably easily from the placenta or umbilical cord shortly after a child is born. The process is safe and can be conducted without harming the baby or mother. The first cord blood transplant was performed in Paris in 1988, and by 2020 more than 40,000 umbilical cord blood (UCB) transplants had been performed worldwide to treat a range of blood cancers and blood and immune disorders.

Given this therapeutic potential in cord blood, since 1993 many healthcare systems throughout the world have developed public cord blood banks to collect and store cord blood for later use as a treatment. Cryopreserved (frozen) cord blood can be kept for more than 20 years, and the fact that it can be made readily available reduces the time required to find the stem cells required for a given treatment. It can take three to four months to find and secure a matching bone marrow donor but can take as little as only two weeks to use UCB stem cells. This can be life-changing for patients who are gravely ill and require a procedure as soon as possible.

A private cord blood banking industry has also developed in many countries, where 'worried well' parents are offered the opportunity to collect their child's cord blood at birth and then store it in case it is needed in the future. This is a fairly expensive 'for profit' service with an up-front fee that tends to be in the low four figures wherever it is available, and additional annual storage costs. Concerns have been expressed about the practice. As a 2020 study conducted in Canada put it: 'Governments have noted the tendency for private banks to oversell the potential for cord blood use, especially in relation to speculative cell therapies not yet supported by clinical evidence.'

Notwithstanding these concerns, by now there are around 5 million units of UCB stored in blood-banks globally, with about 20 per cent of those in public banks and the rest having been stored privately by wealthy parents. To be fair to the private cord blood storage companies and parents who choose to go that way, while they may be selling the merits of cord blood storage on the basis of treatments which are as yet experimental and not yet approved to a certain extent, they can at least point to the fact that there are nearly 80 cord blood treatments which *are* already approved by the US FDA, and a large number of clinical trials continuing to investigate their use for an increasing range of serious diseases and health conditions. You might also forgive parents from making the decision to bank their child's cord blood in places which do not yet have a developed public blood banking system.

This all having been said, it is worth noting that future stem cell treatments may not even need a supply of cord blood. 'Peripheral blood' (or simply 'blood') has also been used for hematopoietic stem cell transplantation nearly as long as cord blood. The approach was first used successfully in 1989 not that long after that first cord blood transplantation procedure took place in Paris.

Peripheral blood, cord blood and bone marrow all have very particular pros and cons in terms of their relative potential efficacy, the speed with which a donor can be found and the risk of rejection which might occur in allogeneic procedures (i.e., ones which use donors rather than the patient's own stem cells). While their use in HSCT, specifically, can treat as many as 80 diseases already as we have seen, most of those diseases are reasonably rare and so the patient population and the number of procedures per annum is still quite small within the context of healthcare overall.

More exciting are the stem cell technologies that may be on their way in the relatively near future. Not only might these have far wider application than those used to date for HSCT, but such technologies may also mean that it may not be too long before clinicians won't even need to harvest stem cells from peripheral blood, cord blood or bone marrow anymore either, even for HSCT procedures. To explain why this may be, it is worth looking quickly at one of the reasons why stem cells have been controversial historically and at another Nobel Prize–winning technology that I would argue kicked that controversy into the proverbial long grass.

Stem cell controversy – ESCs versus iPSCs

Stem cell research was one of the biggest controversies of George W. Bush's presidency back in 2001. In August of that year President Bush banned federal funding of research into stem cells using newly created human embryonic stem cells, or ESCs.

Early stem cell research up to that point had focused on the use of these 'pluripotent' stem cells obtained from 'spare' early-stage human embryos taken from in vitro fertilization clinics. Such embryos came from eggs which had been fertilized but never implanted in an IVF patient's uterus. This use of ESCs involved the destruction of a human embryo, and the practice was widely opposed by religious groups which viewed this as no different from abortion.

A debate raged for several years between those religious groups focused on the sanctity of human life and those on the opposing side of the argument whose position was that the destruction of such an early-stage embryo was morally and ethically justifiable given how many lives could potentially be saved in the long run as a result of the research facilitated by the practice.

In what I think is a rather wonderful example of scientific progress very likely having found a solution to a seemingly intractable problem, in 2006 Professor Shinya Yamanaka (1962–) and his team working in Kyoto, Japan, discovered how to reprogram adult human cells back to an embryonic-like pluripotent state. They called the resulting stem cells 'induced pluripotent stem cells', or iPSCs.

Professor Yamanaka was awarded the Nobel Prize for this work in 2012 alongside another British scientist, Sir John Gurdon (1932–), whose work as long ago as 1962 had provided a crucial foundation for Yamanaka's breakthrough more than 40 years later. Not long behind Yamanaka and his team, in 2007 Professor James Thomson (1958–) and his team working in Madison, Wisconsin, were the first to describe the isolation of human iPSCs.

The discovery and development of iPSCs has led to extraordinary progress in many areas of medicine and, I would argue, largely side-stepped the highly problematic controversy around the use of embryonic stem cells which slowed stem cell research meaningfully at the turn of the century.

Challenges certainly remain. This is extremely complicated science. iPSCs are hard to manufacture and manipulate and they are still very expensive as a result. There are also meaningful safety concerns around their use which must be addressed before any treatments can be approved for widespread use. As a 2020 blog post for Cell Guidance Systems puts it: 'iPSC therapeutics still have much room for improvement on cost, efficacy and safety before they can hope to enter the medical mainstream.' But there is a good chance that these challenges will be overcome in the fullness of time and quite likely sooner than many fear. To a great extent, progress in this area can be exponential in terms of processing power and the tools and analytics available to scientists and clinicians, for example, and also when it comes to the ability of innovative firms to drive significant cost reduction over time, just as was achieved by Oxford Biomedica in the manufacture of the Novartis CAR-T therapy Kymriah we looked at earlier.

The other reason iPSC and indeed other stem cell treatments may deliver on their promise in the relatively near future has to do with 'convergence' and with the fact that rapid forward progress is being made in a number of key related technologies. Arguably, foremost among those are the application of artificial intelligence and machine learning to the science and, related to that, the development of another Nobel Prize-winning technology called CRISPR.

CRISPR and gene editing

In 2020 Jennifer Doudna (1964–) and Emmanuelle Charpentier (1968–) were awarded the Nobel Prize in Chemistry 'for the development of a method for genome editing'. That method is called CRISPR, which is a welcome abbreviation of 'clustered regularly interspaced short palindromic repeats'. In 2018 Doudna and her co-author, Samuel Sternberg, published their book on the subject: *A Crack in Creation: The New Power to Control Evolution*. A UK *Guardian* review of the book by Peter Forbes described CRISPR as '[p]robably the greatest biological breakthrough since that of Francis Crick, James Watson and Rosalind Franklin'. Doudna's biographer, Walter Isaacson, describes CRISPR as nothing less than 'the most important advance of our era'.

Decades from now there is a real chance that we may remember the discovery and development of CRISPR as a step-change in human progress, as hyperbolic as that may sound. CRISPR is another one of the technologies which underpin the entire thesis of this book. In common with most other technologies that win scientific Nobel Prizes, it is certainly complicated. For an explanation of how it works and how it came about, it is well worth reading Doudna and Sternberg's book and also Walter Isaacson's biography of Jennifer Doudna, *The Code Breaker: Jennifer Doudna, Gene Editing and the Future of the Human Race.*

As the term 'gene editing' suggests, at the simplest level CRISPR enables scientists to change or 'edit' a given organism's genome. Using CRISPR, scientists can make extremely accurate changes to DNA whether that be in a plant, fungus, animal or human being. They can therefore correct or repair mutated genes which cause disease, for example. In the laboratory the technology has already been used to correct the genetic mutations responsible for cystic fibrosis, sickle-cell disease, some forms of blindness and severe combined immunodeficiency.

In the long run, the technology has the potential to deliver functional cures for *any* genetic disease where we understand and can correct the genetic mutation that causes that disease. The list of such diseases is long – more than 10,000 human diseases are caused by a defect in a single gene – and these include genetically inherited blindness, cystic fibrosis, muscular dystrophy, blood disorders such as beta-thalassemia and sickle-cell disease, and a great deal else besides.

CRISPR will certainly have a role to play in fighting cancer, too. The technology is already being used to help engineer the T-cells we looked at above, for example. US company CRISPR Therapeutics is already 'developing off-the-shelf, gene-edited T cell therapies using CRISPR, with two candidates in clinical trials'. In the longer term we could potentially even use the technology to genetically edit cancer itself. Researchers have already identified the genetic factors involved in cancer metastasis, for example. Future CRISPR treatments could very well simply switch off cancer's ability to metastasize. This idea is particularly exciting when you consider that somewhere between two-thirds and 90 per cent of cancer deaths are caused by metastases.

Research scientists are also working on functional cures for HIV, COVID-19, influenza and the common cold using CRISPR. Although it was relatively early-stage research, in June 2021 Philip Santangelo and his team working at the Georgia Institute of Technology published a study in the journal *Nature Biotechnology* demonstrating an ability to slow and even stop the replication of the SARS-CoV-2 virus in mice. According to a 'NIH Director's Blog' post, the researchers noted that:

> this approach has potential to work against the vast majority—99 percent—of the flu strains that have circulated around the world over the last century. It also should be equally effective against the new and more contagious variants of SARS-CoV-2 now circulating around the globe.

Because it involves editing the very genetic makeup of life itself, CRISPR has potential applications almost without limit. In agriculture it can produce disease-resistant, hardier crops which can grow more quickly and require less nutrition and irrigation. We could engineer fruits and vegetables to taste better and even ripen more slowly, thus extending shelf-life and significantly improving our agricultural productivity and food supply chains as a result. There are even serious scientists working on resurrecting extinct species such as the woolly mammoth (*Mammuthus primigenius*) using the technology.

CRISPR could be used to make our food healthier and help us combat disease through improved nutrition. In fact, a key foundational piece of research which led to CRISPRs discovery was conducted by two scientists at the Danish company Danisco, Philippe Horvath and Rodolphe Barrangou, who were focused on improving the manufacturing processes for yoghurt at the time.

The technology could change the colour of our fruits and vegetables and potentially even make them glow by inserting DNA from a particular jellyfish, as wildly science fiction as that sounds. While I doubt that any of us are going to demand glow-in-the-dark apples, carrots (or pets for that matter) any time soon, that same technology has much more viable application in a range of sophisticated diagnostic technologies using 'optogenetics' and 'fluorescent proteins'.

Editing the very stuff of life is not without controversy, of course, and this becomes particularly pronounced when it comes to gene editing in human beings and most particularly of the human 'germline' – that portion of our genetic make-up which is heritable. CRISPR could be used to make changes to organisms that would be inherited by their progeny, that is, they would become permanent. This is highly controversial. Doudna and Sternberg raise the question of whether it is appropriate to use gene editing in unborn children to decrease their chances of developing heart disease,

Alzheimer's, diabetes, or cancer. Or, to bestow unborn children with advantageous characteristics such as enhanced strength and cognitive abilities, or altering physical attributes like eye and hair colour.

They then go on to say: 'For the first time ever, we possess the ability to edit not only the DNA of every living human but also the DNA of future generations – in essence, to direct the evolution of our own species.'

Understandably, this extraordinary potential power raises significant ethical and moral concerns as we grapple with our newfound ability potentially to 'play God'. It also raises the unsavoury spectre of a new kind of 'eugenics'. A 2021 article in *Scientific American* entitled 'The Dark Side of CRISPR' suggested that CRISPR's 'potential ability to "fix" people at the genetic level is a threat to those who are judged by society to be biologically inferior'.

The overriding moral question will be around where society chooses to draw the line between repairing disease and enabling as many people as possible to live healthy lives free from pain and distress, and the risk that the technology veers too far in the direction of 'upgrading' future generations of humankind – which could be problematic for any number of reasons.

This is particularly the case if we consider that the wealthiest in society will be those most likely to wield the power to upgrade themselves and their offspring. It wouldn't take too many generations before such biological inequality significantly reinforced and supercharged the economic inequality that exists today already. Perhaps even worse, in the context of nation states, countries best able to leverage such tools for the benefit of a sufficiently large proportion of their population would have a significant advantage over those further behind technologically both economically and no doubt militarily, too. There are also valid concerns over the use of CRISPR technology as a biological weapon of some kind. The US 'Worldwide Threat Assessment' presented by American intelligence agencies to the Senate Armed Services Committee has already included genome editing as one of six potential weapons of mass destruction.

This is a hugely complicated and nuanced debate. As Doudna and Sternberg themselves write:

> Some people view any form of genetic manipulation as heinous, a perverse violation of the sacred laws of nature and the dignity of life. Others see the genome simply as software – something we can fix, clean, update and upgrade – and argue that leaving human beings at the mercy of faulty genetics is not only irrational, but immoral.

What is thoroughly encouraging here, however, is her elucidation of the extent to which leading scientists working in these fields have been willing to impose real discipline on themselves voluntarily, and for several decades. Even before the development of CRISPR, as long ago as the early 1970s, scientists working on the emerging technology of recombinant DNA we looked at earlier, realized their work could have a number of unpredictable and dangerous consequences. Doudna cites the example of the actions of Paul Berg, a Stanford biochemist (who went on to win the Nobel Prize) and his colleagues and peers who asked the US National Academy of Sciences to establish a committee to formally investigate and monitor the development of recombinant DNA. The Committee on Recombinant DNA Molecules first met at the Massachusetts Institute of Technology in 1974 and, following that meeting, produced a report highlighting the potential biohazards of the technology. The report called for a worldwide moratorium on a whole class of experiments that the committee feared might be particularly dangerous. As Doudna explains: 'It was one of the first times that scientists had voluntarily refrained from conducting a whole class of experiments in the absence of any regulatory or governmental sanctions.'

Fast-forward 40 years to late 2014 and Jennifer Doudna and *her* colleagues and peers took a similar approach. In January 2015 they held the first Innovative Genomics Institute Forum on Bioethics to discuss the moral and ethical implications of CRISPR. The meeting was attended by leading scientists, including Paul Berg himself, clinicians and professors of law and bioethics. Other researchers who were not able to attend were nevertheless included in the discussion thereafter and contributed to an article that was published in *Science* after the meeting titled 'A prudent path forward for genomic engineering and germline gene modification'.

Published in March 2015, the article asked fellow scientists to 'hit the pause button' when it came to using the technology for anything to do with altering the human genome and called for an international meeting 'to ensure that all the relevant safety and ethical implications could be openly and transparently discussed'. The authors called for the inclusion of diverse stakeholders such as religious leaders, champions of patient and disability rights, experts in social sciences, regulatory bodies, and government agencies.

Happily, the piece received a great deal of attention, including a front-page story in the *New York Times* and coverage from National Public Radio and the *Boston Globe*. Within a few months the American Society of Gene and Cell Therapy, the International Society for Stem Cell Research and the White House's Office of Science and Technology Policy had all agreed that there should be a moratorium on any use of gene editing to alter the human germline.

In the time since, there has been a reassuring focus on moving carefully with the technology from many of the key stakeholders globally. That having been said, there are many in the scientific and medical community who believe that, 'when all safety, efficacy and governance needs are met, there may be morally acceptable uses of this technology in human reproduction'. As one of the participants put it in that first January 2015 meeting of the IGI Forum on Bioethics: 'Someday we may consider it unethical *not* to use germline editing to alleviate human suffering.' As with every huge scientific leap forward, 'With great power comes great responsibility.' I think by now it is clear that most of the leading players in this field globally are treading carefully.

As an illustration of this reality, it is worth considering the case of the Chinese scientist Professor He Jiankui. In November 2018, Professor He scandalized the world of science when he announced at a conference in Hong Kong that he had edited the genes of two twin baby girls, Lulu and Nana. These were the first children in the world to have been gene-edited. His announcement was met with wide condemnation from all over the world. He was sacked by his university in Shenzhen and sentenced to three years in prison.

Miracle cures, today…

In the meantime, there is much that CRISPR can deliver without the need to alter the *heritable* human gene line. In terms of the development of therapeutics, the first viable use of the technology is to look to repair the defective gene in one of those 10,000 diseases that are caused by only one such gene.

With respect to that, at the time of writing the first CRISPR treatment has just been approved. In the last quarter of 2023, the UK MHRA and US FDA approved CASGEVY, the first ever CRISPR-based gene-editing therapy for the treatment of sickle-cell disease.

Sickle-cell disease (SCD) is an inherited condition caused by a genetic mutation which impairs the body's ability to make normal haemoglobin – the key protein in red blood cells. The mutated haemoglobin can make the patient's blood cells rigid and sickle shaped. These cells can then get stuck in the body's blood vessels with all sorts of awful side effects. The main symptoms include episodes of acute pain. Sufferers face an increased risk of heart attack, blindness, bone damage and severe organ damage to their liver, kidney, lungs and heart, need regular hospital visits and morphine infusions to deal with their pain episodes, and have a much shorter life expectancy than the rest of the population.

Jimi Olaghere was one of the first few patients enrolled in the CASGEVY trial. As he explains about his condition: 'It is like shards of glass flowing through your veins or someone taking a hammer to your joints. You wake up in the morning with pain and you go to bed with pain.' In September 2020, Jimi had his own CRISPR-engineered red blood cells put back into his body. Two weeks later, as reported by BBC journalist James Gallagher, he 'emerged as a completely new person' having woken up with no pain. As he put it: 'I'm living life as a new person. [...] Just being able to take a walk with my son, that's something I thought I was going to miss out on.' Dr Haydar Frangoul, the clinician who treated Jimi in the trial, elaborates:

> [T]he data on the first seven patients has been nothing short of amazing. What we are seeing is patients are going back to their normal life, none have required admission to hospital or doctor visits because of sickle cell related complications.

CASGEVY is the first 'miracle-cure' created by the CRISPR technology just over a decade after Jennifer Doudna and her colleagues first published their breakthrough discovery. This is a cure for one of the more common genetic diseases found around the world; there are an estimated 300,000 infants born with it globally every year.

In the next few years, CRISPR may be able to tackle many more such genetically inherited diseases and deliver a great deal more thereafter, both in terms of therapeutics and potentially in terms of revolutionizing food production and food security, too.

'Smaller large molecules'/protein scaffolds

Another source of excitement for the future trajectory of the biotech industry concerns the development of what we might call 'smaller large molecules' (as it were). In the last few chapters we have referenced large molecules, otherwise known as 'biopharmaceuticals' or 'biologics', several times. They are the truest of 'biotech' drugs, and some way more than $200 billion worth of such drugs were sold in 2022.

Notwithstanding this significant value creation and the ability of such drugs to deliver innovative treatments for a wide range of diseases, they have a number of inherent drawbacks as already mentioned in passing. To elaborate in a bit more detail: large molecules such as proteins and monoclonal antibodies are some way more complex to synthesize and manufacture than small molecules. This makes them more expensive and can also introduce the problem of batch-to-batch variability for pharmaceutical companies. Regulators insist that the manufacturing process for any given drug must produce *precisely* the same product every time – that is to say, the specific compound they have approved for use. This can be a real challenge when it comes to the manufacture of immensely complicated large molecule drugs.

Similarly, large molecules are prone to degradation and instability. This makes them more challenging to store and transport and can limit their shelf life, injecting still more cost into their use. They also tend to be less 'bioavailable' than small molecules. This can make it harder for them to reach their target site in the body and to achieve the desired therapeutic effect when they get there. One of the challenges with large molecule cancer drugs, for example, has been that they have often struggled to get into the tumour microenvironment because they are simply too big. This is assumed to be one of the reasons that many of the new immune checkpoint inhibitor drugs can have relatively low response rates in patients. Getting past these problems can require the development of specialized delivery systems or formulations, adding yet more expense and complexity to a given treatment.

Lastly, their molecular size may also trigger an immune response in the body, which can reduce their effectiveness or lead to adverse reactions.

Given all of the above, many innovative biotech companies are working to develop *smaller* biologic drug platforms which might have a chance of mitigating these challenges or even of avoiding them altogether. These are sometimes called 'protein scaffolds'.

In Belgium a company called Ablynx, owned by the French pharmaceutical giant Sanofi since 2018, has developed something called a Nanobody. This is a kind of specialized miniature antibody which was discovered in llamas, alpacas and other camelid species in 1989. Nanobodies are roughly one-tenth the size of a 'normal' human antibody. Because of their small size, they can get around many of the problems associated with the existing class of large molecules given above. They are more robust and stable and can be more easily and quickly produced as a result. They are also potentially easier to administer, possibly even in a simple oral pill for example.

A number of small Nanobody molecules can also be strung together to create what are called multivalent Nanobodies. In the fullness of time, this could mean that a multivalent Nanobody could help the body's immune system fight cancer by creating a bridge between a tumour cell and an immune cell by attaching to several sites on each.

Sanofi and Ablynx are working on Nanobody treatments for a wide range of therapeutic areas including inflammation, haematology, cancer and rare diseases. On the other side of the world, a tiny Australian company called AdAlta is developing its own 'i-body' platform. As the website for the Australian Trade and Investment Commission (Austrade) explains, it has 'taken an antigen-binding domain found in wobbegong shark blood and transformed it into a humanised single domain antibody that can potentially change treatments for diseases such as fibrosis and cancer.'

Moving away from llamas and sharks, in the UK, British company Avacta plc has its own similar proprietary technology called an Affimer. Just as with the Nanobody and i-body, the Affimer is a protein which is smaller than an antibody. It is also a *human* protein, which likely reduces the risk of an immune response in a human patient as against other such technologies which re-engineer animal proteins. Just like 'multivalent' Nanobodies, Affimers can also address more than one therapeutic target. They are relatively cheap to manufacture, robust, stable and highly soluble.

Avacta's other main technology is called the pre|CISION tumour-targeting platform. This can be used to modify an existing chemotherapy treatment so that the active drug is only released once the molecule has reached tumour tissue. Avacta began a Phase I clinical trial of this platform in the summer of 2021. Although this is an early stage trial the results so far have been encouraging. In the fullness of time this technology may be able to do nothing less consequential than deliver a chemotherapy treatment without side effects. Longer-term, the combination of the pre|CISION technology with the Affimer may be able to deliver a drug molecule which can be taken as a pill, travel through the body without causing any side effects and then deliver a powerful cocktail of a cancer-killing chemotherapy and an immunotherapy payload direct to a tumour.

As I said earlier in the book, although this outcome is still some years away and there is certainly scope for disappointment along the way, the mere fact that such technologies are in development is surely more than a little encouraging.

Much of this is early-stage stuff but there is a great deal of potential in this technology overall and there could be a good deal of excitement to come out of one or other of these companies or from one of many others working in the field in the next few years.

The importance of diagnostic and analytical tools: PCR and NGS

When we consider the extraordinary and exciting developments we have covered in this section of the book, from recombinant DNA to messenger RNA, monoclonal antibodies, gene and cell therapy, stem cells, CRISPR and new, smaller protein-based biologic therapies, another crucial development that had to happen in parallel for any of that progress to be possible has been the arrival of sophisticated diagnostic technologies and analytical platforms.

We have already seen how important the emergence of industrial chemistry was in the nineteenth century for the development of the pharmaceutical industry and of small molecule drugs in the first instance. Within a few decades, a raft of new innovative analytical techniques such as X-ray crystallography, paper chromatography, ultraviolet spectroscopy and many others were key to the progress made and to the work which enabled Watson, Crick, Wilkins and Franklin to complete their elucidation of the structure of DNA, for example.

When it comes to the development of the biotech industry in particular, in the last few decades there are two analytical technologies that have been crucial for progress: polymerase chain reaction (PCR) and next-generation sequencing (NGS).

Polymerase chain reaction (PCR)

PCR was developed in the 1980s by the American biochemist Kary Mullis. Mullis was awarded the Nobel Prize in Chemistry in 1993 for his invention of the technology. PCR enables scientists to take a very small amount of DNA and amplify it to produce enough for analysis. This was critical for the development of the biotech industry and in other fields such as forensic science and even archaeology and palaeontology.

PCR is used to detect and analyse cancer and pathogens and to diagnose infectious disease. It is the technology used for DNA profiling in criminal investigations and for testing parentage, as seen so often in Hollywood films and television dramas. Because it can amplify a tiny quantity of DNA, it has also been used successfully on long-extinct animals, such as mammoth fossils more than 1 million years old, and on ancient or very old human DNA such as that found in Egyptian mummies, the bones of Neanderthals, or in historical figures including the Russian Romanov Tsar Nicholas II and the English King Richard III.

Since the 1980s, PCR has been an exceptionally important and powerful technology for the development of the biotech and many other industries. It is highly sensitive and reasonably inexpensive which has meant that it has been widely adopted and is a familiar technology for scientists all over the world. This has been supercharged in recent years with the arrival of battery-powered, portable PCR machines. As with most technologies, the first generation of PCR machines were fairly large, requiring space in a lab or hospital and mains power. This also meant that DNA samples collected remotely would need to be shipped to that lab or hospital even if they may have been acquired hundreds or even thousands of miles away in a refugee camp suffering a disease outbreak or on a farm for livestock testing for example.

The emergence of small, lightweight, portable PCR units that are sufficiently robust and accurate in the field has enabled scientists to take the technology where it can be most needed and to widen the potential applications for its use: in food and beverage production, water and environmental testing, agriculture, and for remote or drive-through sites as occurred during the COVID pandemic of course. There is even a MiniPCR machine on the International Space Station today, which would have been unthinkable not that long ago.

While PCR is a fantastically useful and versatile analytical and diagnostic technology, it has certain drawbacks. It can only analyse a relatively small number of genetic targets simultaneously, and it can also only provide information on whether a given DNA sequence is present or absent. That is to say that a scientist has to decide what specific DNA or RNA sequence they are looking for – which disease or bacteria for example – and the PCR machine can then indicate whether it is there or not. There can also be a risk of 'false positives' if the DNA sample is contaminated and of 'false negatives' if there is too little target DNA in a sample or if that DNA has not been effectively amplified such that the PCR machine can't detect it.

Next-generation sequencing (NGS)

Happily, most of these drawbacks have been addressed by the emergence of another key technology roughly since the early 2000s: next-generation sequencing, or NGS.

Earlier in the book we looked at the incredible fall in the cost and time required to sequence a genome – to read the code of a given sample of DNA or RNA. NGS is the technology that enabled us to do this. It is not an exaggeration to say that it has revolutionized all sorts of research. The technology has been developing roughly since the turn of the century. It has been honed and evolved since then over several iterations to become arguably one of the most important technologies of the modern age and certainly for the biotech industry.

NGS, also known as 'high-throughput sequencing', enables scientists to analyse a very large amount of genetic information quickly and reasonably cost-effectively. This has wide application in research and clinically. The technology enables scientists to read the detailed genetic code of whatever they might want to look at. This could be a person, animal, plant, fungus, bacterium, virus or tumour, for example. NGS can give us rich information about specific cancer variants and the state of a patient's disease. It is also a crucial analytical tool for studying the immense complexity of the microbiome and virome which we looked at earlier in the book.

Of course, being able to read and document genetic code is only one piece of the jigsaw puzzle. Researchers and clinicians will also need to understand the biological or medical meaning implied by that data. Reading someone's genetic code is one thing. Recognizing the fact that they have a genetic mutation that causes or increases the risk of a given disease is another. Similarly, securing the accurate genetic code of a patient's tumour isn't particularly useful unless you can say something substantive about what that code or gene sequence implies for the patient: is their tumour benign or aggressively malignant, for example? If it is malignant, what treatment is likely to be the best option based on the detailed genetic make-up of that tumour?

This is a key component of the technology. The biotech and medical device companies which provide the machines which can read the genetic code of a DNA or RNA sample also provide access to a large and constantly growing database of much of the genomic research that has been conducted to date all over the world so that researchers are able to draw substantive conclusions from their work.

The US company Illumina is the largest player in the NGS market by some margin, with an estimated 80 per cent market share. It is the 'Intel inside' of the gene sequencing market. Its machines come with its BaseSpace Correlation Engine software. This gives scientists access to a vast and growing quantity of existing data and research, including more than 20,000 existing genomic studies. This is astonishingly complex stuff involving an eye-watering amount of data. The human genome is made up of approximately 3.2 billion 'base pairs' of DNA but most of the trillions of cells in our bodies have two copies of your genome, implying that there are around 6.4 billion 'letters' in the genetic code in our cells.

When you consider that many millions of NGS sequences have been run since the technology was invented in the mid-2000s, each with billions of underlying data-points, the sheer scale of the volume and complexity of sequencing data generated by NGS technologies and the computational power required to interrogate that data and draw useful conclusions becomes clear. Most of us are familiar with the idea of a megabyte (MB) or gigabyte (GB) when it comes to computer memory. A terabyte (TB) is 1,000 gigabytes, a petabyte (PB) is 1,000 terabytes and an exabyte (EB) is 1,000 petabytes (next come zettabytes and yottabytes for what it's worth). As long ago as 2019 it was estimated that the total volume of NGS data generated up to that point was around 200 exabytes (EB) and that this volume was doubling approximately every seven months. It's important to note that these estimates are constantly changing as more data is generated and analysed, and as new technologies and methods are developed to handle and process the data, but the overriding point is that this is an extraordinary quantity and quality of data and growing day by day, hour by hour, a great deal of which is in the public domain and accessible by scientists everywhere.

The digital output from NGS machines is by no means the only large dataset in the public domain. For some years, medical images have also been collected and made available to academics and clinicians to use in teaching and for reference. The US National Library of Medicine launched MedPix in 2016, for example. It is a free open-access database of approaching

60,000 (and growing) medical images from more than 12,000 patients. It is by no means the only such database available. Today researchers and clinicians all over the world can access historical datasets which include simple pictures (of a wound for example), CT and MRI scans, X-rays and a great deal else besides. Such datasets include supporting data related to the images which would include patient outcomes, treatment details, genomic information and expert analysis from clinicians or scientists from a clinical trial, for example.

It is worth taking a moment to reflect on just how incredible this technology is and also on how wonderfully collaborative its use. None of this would be possible without (extremely) big data, the phenomenal processing power available to us today after a century or more of Moore's law and ubiquitous high-speed broadband internet connections. The fact that scientists, clinicians and regulatory bodies all over the world are willing and able to publish and upload many of their findings for use by their peers all over the world to engender yet more forward progress is also key, and philosophically very far away from what has so often been the case in the past. This has also only happened relatively recently and, again, is developing exponentially.

I have already made the point that German, Japanese, US and British scientists worked in isolation and secrecy during the Second World War for obvious reasons. Not only have processing power, memory and communications technology increased exponentially in recent decades, so has the willingness of human beings to work together with such technologies on so many of our most intractable problems and not just in the healthcare setting.

The role of AI and machine learning

Yet another technology that has been developing in parallel at an exponential rate, and has entered the contemporary zeitgeist most particularly at the time of writing with the arrival of high-profile AI 'chatbot' ChatGPT, is that of artificial intelligence (AI) and machine learning (ML).

With trillions of data points and exabytes of information being generated by many thousands of research scientists all over the world, the task of extruding meaningful conclusions which can move us forward scientifically has become increasingly and massively more complex. Happily, the rapid evolution of AI and ML at roughly the same time as the emergence of NGS and other similar technologies has meant that we are able to address that complexity more and more. This is yet another area where science fact increasingly really does look like science fiction and where what we are capable of doing or aspiring to do is truly extraordinary.

AI and ML can be used to improve drug discovery and drug design, to optimize clinical trials before they begin, to repurpose existing drug molecules, to hone manufacturing processes, to speed up diagnosis from medical imaging and a great deal else besides. Much of this can and will strip significant cost out of the industry, too.

The British company Exscientia is one of the leaders in the field. As Andrew Hopkins, its CEO, explains, AI gives the company 'the ability to search a much vaster chemical space than traditional processes could hope to handle'. AI can look for patterns that are too complicated for a human to recognize. The same can be true when looking at medical images. Various computer-aided detection (CAD) systems have been approved by the FDA already in this area. They can be used in conjunction with mammography to help radiologists detect breast cancer, for example. Such systems use AI algorithms to analyse images and highlight areas that may be indicative of cancer. Similar approaches have been approved for use in automated retinal image analysis to screen for diabetic retinopathy and refer patients to ophthalmologists for further evaluation if required, to interrogate CT scans to help clinicians work out whether a lung nodule is likely to be cancerous or benign, and to detect and classify brain haemorrhages.

Another exciting application for the technology is in repurposing existing compounds. Precision Life is another highly innovative British company in the field. In June 2022, it published research in the journal *Cell Patterns* which analysed the drug pipelines of 177 biopharma companies using ML. Its goal was to find other potential uses for *existing* drugs. This broad approach had yielded some success when addressing coronavirus during the pandemic, for example. As Precision Life's CEO, Dr Steve Gardner, explained at the time, 'The successful global effort to find effective interventions for the most severe COVID-19 patients demonstrated the value of reusing current drugs in new indications.' Precision Life and others like it wanted to go further and undertake a similar analysis for a much wider range of drugs and diseases. If we can find other uses for existing drugs, this could save the biotech industry and healthcare systems many billions of dollars and, according to Gardner, 'provide a faster, cheaper and derisked route to the approval of new therapies, with major benefits to patients'.

Using sophisticated 'combinatorial analytics' from a huge amount of data, Precision Life's work has identified 477 'repositioning opportunities' across no fewer than 35 diseases. That is 477 drug compounds that could potentially be used to do something more than what they currently do. This is only one example of what AI and ML can do. In future, this sort of work is likely to yield significant benefits for the industry and for sick patients.

It isn't just relatively new, small, innovative companies working in this area either. Apple, Google, IBM, Microsoft and plenty of other large companies are also significant players. Google Health is particularly focused on genomic analysis and AI-enabled imaging and diagnostics. Microsoft has a multi-year strategic alliance with companies such as Novartis and with the British gene and cell therapy company Oxford Biomedica, as we have seen. Apple would like to leverage more than 1.5 billion iPhone users globally to collect and analyse medical data which could be of enormous value to researchers everywhere in the years ahead. Apple has also shipped more than 200 million Apple Watches since they first came to market and is fairly ambitious about the technology it would like to embed in that product in future when it comes to healthcare. Apple and other firms like it will drive growth in various wearable technologies in the years to come which will provide even more useful, rich data for any number of biotech industry initiatives using AI. All of this activity is another example of the importance of 'convergence' as a theme overall.

AI and ML will have an exciting role to play in the future development of the biotech industry. Many of those involved in the research believe that eventually it may even be possible to develop drugs '100 per cent *in silico*' – that is to say, by just using computers without the intervention of chemists. As the global pharmatech company Exscientia puts it on the landing page of its website: 'In the future all drugs will be designed with AI.'

Another possible outcome from the realms of science-fiction, still some years away, could be to do away with the need for animal testing – once we have sufficient confidence in the accuracy of the computer models. A 2019 study using a database of 10,000 chemicals and 800,000 studies was able to beat animal testing at predicting toxicity already, for example. AI can already be used today to reduce the number of animals needed for testing by screening molecules *in silico* and optimizing the design of experiments as a result.

There is also the promise of increasingly personalized medicine. In the not too distant future it may be possible for a diagnostic technology to take a sample from a patient (saliva, blood or stool, for example), draw an extraordinarily accurate conclusion about their condition very quickly, and then even design and produce a genetically and biologically personalized treatment on the spot, or within a few hours or days, as the technology develops.

In summary

In this chapter we've seen how the confluence of several key technologies is impacting both the diagnosis and treatment of disease. The promise of the kind of 'precision medicine' outlined above is driven by the potent combination of NGS sequencing, the advance of gene editing with techniques such as CRISPR so that scientists can actually change DNA sequences by introducing or correcting genetic mutations, and the use of enormously sophisticated AI and ML systems to understand and inform all of the above.

In the following chapter we turn our attention away from human health to the world of biotech 'without' us. It is important to recognize that these two 'worlds' are intimately and inextricably connected. Human beings constitute just one element within a broader ecosystem, and the vitality of our environment is intricately linked to our own wellbeing. As we shall see, here too biotech has a key role to play.

10

Biotech 'without' us

Clean power, agriculture, bioremediation and processing power

So far in this part of the book we have focused primarily on the role biotech has to play when it comes to healthcare and on the development of therapeutic treatments in the main. Crucially, many of these technologies have a role to play 'without' us as well as 'within' us, or will do in the near future, including CRISPR and the use of NGS, AI and machine learning in particular. This is arguably just as exciting and likely even more consequential for our species overall.

Clean power technologies

As an example, the industry is already having a meaningful impact on clean power generation. One area of focus here is on innovative biofuels derived from organic matter such as plants or algae. Biotech companies can use a range of technologies to enhance biomass conversion into energy and widen the range of sustainable feedstock options available. Fuel derived from algae is a particularly exciting area of research with bio-derived diesel and jet-fuel both touted as being scientifically possible. Although we are certainly several years and a few billion dollars away from such technologies making a meaningful dent on fossil-fuel consumption, exponentials will have a role to play as we are increasingly able to genetically engineer plants and microorganisms for higher yields and improved energy content.

Biotechnologies have already driven the development of the biogas industry where engineered microorganisms and enzymes are applied to agricultural and industrial waste to produce methane, a valuable source of renewable energy. The International Energy Agency (IEA) estimated that biogas accounted for 2 per cent of total renewable energy consumption worldwide in 2019 but noted that there is much potential inherent in the technology. The European Biogas Association has estimated that biogas could provide 5–10 per cent of Europe's total energy demand by 2050. Such estimates could prove to be conservative given the pace at which such technologies are developing.

At an earlier stage, but also potentially exciting, is what biotechnologies may be able to contribute to the development of the solar industry. 'Biomimicry' is a term used to describe the practice of learning from nature to solve human challenges in areas such as design and manufacturing processes. Biotech researchers are looking to draw inspiration from the natural world to develop materials with improved properties for solar energy applications. By studying natural photosynthetic processes, researchers can develop materials that mimic the efficiency and functionality of photosynthesis in capturing and converting sunlight.

In the fullness of time there may even be scope to produce bio-based solar cells which can be cultured or grown and which will harness solar energy through biologically derived components. This could deliver significant advantages when it comes to the cost, sustainability, and efficiency of solar power overall. There are already bio-coating products which are improving the output of existing solar technologies. Scientists are also working on similar technologies in the realm of carbon capture and storage (CCS) which could have a key role to play in reducing emissions across the energy industry.

Bioremediation

Another exciting application for the industry is that of bioremediation. Nowadays, we are all too familiar with the fact that since the advent of the industrial age we have challenged our environment with any number of non-biodegradable organic and inorganic contaminants, including heavy metals, chemicals, plastics and various effluents. Over time we have developed an increasing understanding of the damage caused as a result: many of these pollutants are carcinogenic and mutagenic to humans and to the animal and plant kingdoms.

The accumulation of heavy metals, for example, can damage our cells and impact our genome and epigenome, increasing our risk of cancer and plenty of other diseases. Elevated levels of pollutants in soils inhibit nutrient absorption and crop growth and reduce the nutritional quality and quantity of our food as a result. There is growing evidence that this is another causal factor in the rise of those various diseases of modernity or modern plagues we looked at earlier in the book.

Bioremediation is a process that uses living organisms, including microorganisms such as bacteria and archaea, plus plants, fungi and enzymes, to degrade, de-toxify or remove contaminants from the environment. Bioremediation technologies can break down or metabolize a range of harmful pollutants, including organic compounds, heavy metals and hazardous chemicals. The approach offers several advantages over traditional remediation techniques. It can be far less expensive, more environmentally sustainable, and often results in complete degradation of contaminants into non-toxic end products.

Various bioremediation technologies have been developing for several decades, roughly from the 1970s onward and the industry was already valued at well north of $100 billion by the end of 2021, but it is only in the relatively recent past that we have developed the analytical tools and expertise in molecular biology and genomics to begin to step-change our understanding of plants and microbial communities and their functional capabilities in this area.

The real promise of the industry lies in the use of a number of complementary technologies together and of a complex cocktail of different microorganisms as well as plants. Today there are an increasing number of companies using technologies such as NGS, AI and machine learning, to address that complexity. Such technologies show great promise, and we are only just getting started given how relatively recently many of the key analytical tools and technologies have arrived.

Agriculture

One example of where new bioremediation technologies show particular promise is in rolling back the damage caused by pollutants to soil fertility. The biotech industry can do a great deal to enhance soil fertility and agricultural output more generally. For centuries farmers all over the world have used traditional techniques such as crop rotation and leaving fields 'fallow' for periods of time to improve soil fertility and agricultural productivity. Increasingly, the biotech industry can accelerate such processes with precisely engineered microbial biofertilizers and biostimulants, over and above repairing agricultural land which has been harmed by contaminants.

Biotech companies are also using advanced techniques to accelerate the development of improved crop varieties. Genetic analysis can identify desirable traits in crops resulting in higher-yielding, disease-resistant, and drought-tolerant plants. Genetically engineered insect-, bacteria-and fungus-resistant crops can reduce the reliance on damaging chemical pesticides minimizing water and soil contamination and enhancing yields. When applied, such approaches can be a great deal more effective and accurate than traditional techniques which have emerged after generations of trial and error.

Perhaps even more exciting than what the industry can deliver when it comes to crop production, however, is the extent to which it is set to revolutionize livestock farming, aquaculture and, by implication, the fishing industry, too. Biotech can play an exciting role in the development and production of meat, fish and seafood substitutes. Such products fall into two broad categories. First, there are meat-like products that are derived from plants or fermentation processes. These have been around for many years but (bio)technology has accelerated their development recently and delivered products that are ever closer to the 'real thing' in terms of look, feel and taste. Arguably, the two leading companies in this field today are US-based Beyond Meat and Impossible Foods. Both companies produce a wide range of plant-based cultivated meat products including chicken, sausage, beef and pork analogues.

Impossible Foods already has a deal with US fast-food chain Burger King to supply a plant-based Whopper burger. Similarly, Beyond Meat has collaborations with Kentucky Fried Chicken, Pizza Hut and Starbucks. Both companies offer a wide range of consumer products too. It is still relatively early days for both companies and for the industry overall both in terms of their commercial trajectory and, related to that, widespread adoption of these sorts of products by consumers. Nevertheless, industry analysts estimate that the market for plant-based meat will grow at anywhere between 25 and 50 per cent per annum for the rest of this decade implying tens of billions of dollars' worth of revenue by 2030 and beyond and far wider consumption of such products than in the past.

Second – and arguably more exciting for meat-lovers who aren't particularly excited by plant-based meat substitutes – is the development of 'cultivated' or 'cultured meat', otherwise described as lab-grown, cell-based, 'no kill' or 'clean' meat. Rather than slaughtering animals to supply meat or attempting to emulate it using various plants, cultivated meat takes real animal cells and grows them in a controlled environment. The holy grail here is to manipulate the cellular structure of the output from that process to deliver an end-product that looks, feels, smells and tastes precisely the same as the real thing. In fact, given it will be genetically the same as meat derived from an animal to all intents and purposes such products *are*, or at least *will be*, the real thing.

The first lab-grown burger caused quite a stir when it was served in 2013, partly because it had cost $330,000 to produce but also because one of the investors in the company which developed it was Google co-founder Sergey Brin. Happily, since then exponentials have driven the price of a cell-cultured burger below $10. This still makes the product expensive to the end consumer and the price is not yet at a level where we can expect wide adoption but it is indicative of the direction of travel and potential.

Mince-based products such as burgers or meat balls are inherently easier to grow in a lab than a steak or fish fillet given how much more complicated the structure of such things are, but here too exponentials are likely to do their job. There are already as many as a hundred companies working on the problem globally. Such companies are also working on lab-grown leather and there are plenty of other products to which such technology can be applied. As one study by McKinsey & Company has put it, success here 'could mean that one day consumers will pay no more for Wagyu beef and bluefin tuna than for chicken nuggets and burgers'.

Far more important, however, will be the fact that this beef and tuna will come from a lab up the road and the impact this will have on the environment, human health and animal welfare. Present-day agricultural practices are one of the leading contributors to environmental degradation, species extinction and climate change. The farming industry has caused 80 per cent of global deforestation, consumes a vast quantity of our steadily diminishing supplies of fresh water and is a major polluter, often with horrific consequences.

Vast quantities of animal and human sewage have been dumped into our environment by the industry in the last century or so, poisoning our land, rivers and oceans with a frightening array of toxins and increasing antibiotic resistance, too, with all that this might imply for our health in future. Our farming practices have also increased the risk of zoonotic diseases transmitted from animals to humans which could be just as bad or even some way worse than COVID-19.

Whatever your views on livestock farming and meat-eating more generally, there is no question that it is also often extremely cruel. Jim Mellon points out that industrial animal farming and its products are frequently extremely cruel. As he then goes on to say: 'Many people are unaware that the food they eat is quite often the result of practices that would disgust them if they were better informed.' If our current agricultural practices continue on the same trajectory of the last century or so, the outlook for us all is likely to be incredibly bleak given the impact this will have on our environment, human health, pollution and biodiversity and particularly for the poorest billion or more of our species in those areas of the developing world most impacted by climate change. This is arguably an area of human endeavour where exponential solutions really will need to outrun exponential problems if we are to avert disaster. Unquestionably, we need nothing short of a revolution in our global food system.

If we can pull off just such a second agricultural revolution, however, there is much cause for hope. As British writer George Monbiot has put it: 'The most secure and effective way of removing carbon from the atmosphere is to reduce the amount of land we need for farming, and rewild the land we spare.' As he goes on to explain: '[if] the land now occupied by livestock were rewilded, the carbon draw down from the atmosphere by recovering ecosystems would be equivalent to all the world's fossil fuel emissions from the previous sixteen years.' This one change could make the difference between success and failure with respect to preventing the 1.5°C rise in global temperatures which many fear could push the planet to some kind of terminal tipping point.

Similarly, the industrial fishing industry is quite obviously the chief cause of the despoilment of our oceans and seas, an immensely concerning loss of marine life and biodiversity and also of marine plastic pollution.

The upside of lab-grown meat, fish and seafood cannot be understated given how high the stakes and it is here that the biotech industry could be the 'silver bullet' required. The impact the biotech industry should have on agriculture may well be the most important of all, even over and above curing cancer or so many other intractable diseases of our modern era.

Packaging

Similarly impactful is what the industry may be able to do with packaging and to reduce our reliance on single-use plastics. By leveraging advances in biotechnology and materials science, companies all over the world are exploring alternative materials derived from renewable resources that can provide similar functionality to traditional plastics but with a significantly reduced environmental impact.

Biotech companies are researching and developing biodegradable polymers derived from natural sources such as plant starches, cellulose and algae and the use of microorganisms to produce alternative materials. Such products can break down into harmless natural components, reducing the persistence of plastic waste in the environment and enzymes derived from biotech research are being explored for their ability to break down existing plastic waste more efficiently too.

Processing power

Another important contribution the industry can make is in the development of processing power and with the future trajectory of Moore's law. Today our conventional silicon-based microchips and processors are bumping up against the more or less intractable limits of physics and Moore's law is becoming increasingly difficult to sustain.

As transistors have shrunk in size and more components are packed onto each chip, space, heat generation and power consumption have all become major challenges for manufacturers. Similarly, as transistor density has increased, the speed at which data can be transferred between different components on a chip or across different chips is an increasingly problematic bottleneck.

Such challenges are driving the exploration of alternative computing technologies, including biologically- and DNA-based computing which could very well revolutionize computing technology and sustain the continuation of Moore's law well into the future. Such technologies leverage the unique properties of biological molecules, such as DNA, to perform computational tasks. At the simplest level, because DNA has four nucleotide bases (adenine, thymine, cytosine, and guanine), DNA processing and storage can be exponentially more powerful than the binary technologies used to date.

In the fullness of time, these approaches could also be far less power-hungry than existing processing and storage technologies and take up far less space than the racks of many millions of computer servers hoovering up electricity all around the world. DNA storage in a living organism may even need no power at all and could be a stable data-storage medium for thousands of years. The potential applications of biologically and DNA-based computers are vast.

Companies working in the field include the likes of Microsoft, Google, Illumina and Western Digital and less well-known, smaller, innovative businesses such as Ginkgo Bioworks, Twist Bioscience and DNA Script. It will be exciting to see how the field develops in the next few years.

In summary

This chapter has necessarily been something of a whistle-stop tour through what biotech is likely to be able to do 'without us'. Each one of the industries and applications covered over the last few pages would merit a book in their own right but I hope this brief canter through them has served to give a flavour for just how exciting some of this technological progress will be.

In the next and final chapter, we will look at one of the most controversial and 'science fiction' ideas to emerge from the industry in recent years: the idea that, in time, age may be nothing more than a disease – and one that may be treatable.

11

Longevity and juvenescence – age is just a disease

Throughout this third part of the book, we've looked at how the biotech industry is impacting the development of novel therapeutics, the extraordinary advance in our key analytical and diagnostic tools, and at the role biotech has to play 'without' us with respect to crucial advances in clean power, bioremediation and, hopefully, in revolutionizing agriculture and computing too.

Arguably, the most 'radical' outcome the industry may deliver in the next few decades is to change our relationship with ageing and death. Earlier in the book I referenced the British anti-ageing scientist, or 'biomedical gerontologist', Aubrey de Grey, who has argued for some years now that the first person to live to 1,000 years old may already have been born. In fact, as he has put it: 'The 1,000-year number is purely a ballpark estimate of the average lifespan – and even then, it's in the context of today's risk of death from causes that don't arise from ageing, so it's almost certainly very conservative.'

De Grey's position, shared by a growing number of other entirely serious scientists, is that ageing is actually a disease and should be treated as such, rather than being seen as a natural and inevitable process. In fact, their view is that ageing is *the* paramount disease of our species. It kills roughly 100,000 people a day, after all, and is the ultimate underlying root cause of most of the other diseases to which we attribute mortality, including cardiovascular and respiratory diseases, cancer, Alzheimer's, dementia and Parkinson's disease, diabetes and fatal diseases of the kidney, liver and gut.

Arguably, one of the highest-profile people working in the field is David Sinclair. Originally from Australia, he is a tenured professor in the Department of Genetics and co-director of the Paul F. Glenn Center for the Biology of Aging at Harvard Medical School. He also jointly heads the Ageing Labs at the University of Sydney. Professor Sinclair has published more than 170 scientific papers, has been a co-inventor on more than 50 patents and involved in the foundation of 14 biotech companies. He is also co-founder and co-chief editor of the scientific journal *Aging*. In 2014 he was included in *Time* magazine's list of the '100 Most Influential People in the World' while still only in his mid-forties, and was listed in '*Time*'s Top 50 in Healthcare' in 2018.

In 2019 Professor Sinclair published his seminal book *Lifespan: Why We Age – and Why We Don't Have To*. In the book he argues that, if we can understand the underlying mechanisms of ageing, we can develop treatments to slow down or even reverse the ageing process. He explains the role of genetics but, more importantly, that of *epigenetics*. This is the study of changes in gene expression that do not involve changes to our *underlying* DNA. Our genes play a significant role in determining our lifespan, but there are also external factors, such as diet and lifestyle, that can influence how our genes are *expressed* and this is likely to be far more important. Sinclair states that 80 percent of your future health is within your control and can be modified, while 20 percent is determined by genetics, which you have limited ability to influence. He goes on to explain, more or less robustly: 'There is no biological law that says we must age. Those who say there is don't know what they're talking about.'

While the code of our DNA is essentially 'fixed', the way that code is read and used at the cellular level in the body is what actually controls our lives, our health and, crucially, the rate at which we age as a result. It is epigenetics and the epigenome which determines which of our genes should be turned on or remain off, for example. The effective functioning of our cells, tissues and organs, and the rate at which we accrue cellular damage (ageing) comes down to the role played by our epigenome far more than our genome.

As a wonderful illustration of the difference between the two and of the power of the epigenome, Professor Sinclair explains that a caterpillar and the butterfly it becomes have precisely the same *genome* but the extraordinary changes that occur during metamorphosis are due to changes in *epigenetic* expression. Crucially, while we basically can't influence our *genome* via lifestyle interventions and the way we behave (although in future we may be able to do so using CRISPR, of course), we can influence our *epigenome*. Doing so can and will slow down the ageing process. Such interventions include calorie restriction, intermittent fasting, good sleep, various kinds of supplementation, both low-intensity and vigorous exercise and, interestingly, cold and heat exposure. Sinclair is another leading scientist espousing the merits of hormetic stress which we looked at in Chapter 6.

One of the key points Professor Sinclair makes in his book, exceptionally aligned with the overall message of this one, is how fast the science is moving. He mentions advanced sequencing machines capable of analyzing millions of genes daily, allied to computers capable of processing trillions of bytes of data at speeds previously unimaginable. Additionally, he underscores the fact that this technological progress builds upon centuries of accumulated knowledge. It is the march of science and the key role played by exponentials that are helping us finally understand something as extraordinarily complicated and multifactorial as ageing and work out what we might do about it. Arguably, the most exciting element of the work being done by Sinclair and others like him is that this isn't just about increasing *lifespan*. Far more important, as we are increasingly able to retard cellular damage, it is *healthspan* which we can increase – the proportion of our lives where we are generally in good health. He argues that it is entirely feasible for individuals currently alive and in good health to live to 100 while maintaining an active and engaged lifestyle comparable to that of today's healthy 50-year-olds.

Sinclair and others like him believe this is possible because, increasingly, we have a much better understanding of what causes ageing and have the tools at our disposal to address those causes more or less directly. We can address cell loss by using stem cells, for example, or help the body to remove dead 'senescent' cells which cause inflammation and, eventually, cancer. We can boost our immune systems so they do a better job at clearing damaging waste products. There is a great deal we can do and, seemingly, much of it is reasonably straightforward (even if much of it is not – not yet at least).

As Sinclair puts it: 'From the looks of it, aging is not going to be that hard to treat, far easier than curing cancer.' This is particularly exciting when we consider that the key, treatable, hallmarks of ageing are the upstream causes of many cancers, likely even most of them, and of so many other more complex health conditions and awful diseases. If we can effectively 'treat' ageing, then the downstream effect will be that there will be vastly less cancer to treat, and Alzheimer's, diabetes and so much else.

This overall approach is likely to be enormously more efficient for us as a species than what Sinclair describes as the 'whackamole' approach to disease that we take today. We spend hundreds of billions on medical treatments which give patients a few more months of life, or a few years for the lucky ones. Often, even if they survive a little longer, those patients will have a pretty poor outcome in terms of quality of life, particularly with legacy treatments such as chemotherapy, radiotherapy and surgery. Even the small minority of the luckiest of them who may enjoy progression-free survival on the best new cancer drugs may only have a few more years with their loved ones and with an acceptable quality of life. Treating age as a disease and succeeding in that endeavour may give many millions of us decades of progression-free survival, robust health and vitality and significantly reduce the risk that we get diseases such as cancer to begin with.

Many of the various 'cures' for ageing that Sinclair and others like him are advocating are also vastly cheaper than our most advanced drugs and, successfully adopted, could save our healthcare systems trillions of dollars as a result. As he puts it: 'Treatments that once cost hundreds of thousands of dollars could be rendered obsolete by pills eventually costing pennies to make.'

Sinclair's views are certainly controversial but the same could be said of Copernicus, Galileo and so many others since. As Mahatma Gandhi famously put it: 'First they ignore you, then they laugh at you, then they fight you, then you win.' It is more than a little exciting to contemplate a future where the ideas of Sinclair and others like him 'win'.

A second-order economic benefit of treating age as a disease and succeeding could be many more trillions of positive impact on our economies as people can work and contribute far longer and need far less brutally expensive healthcare than has been the case in the last century or so since we first started to expand lifespan but without expanding healthspan. As Sinclair puts it: 'As it stands, aging presents a double economic whammy, because adults who get sick stop making money and contributing to society at the same time they start costing a whole lot to keep alive.'

This possibility raises controversial philosophical questions that go to the very root of how we function as a society and even as a species assuming a widespread capacity to combat ageing effectively. What would the world be like if many of us live well past the age of 100? As Sinclair puts it, 'such a transition cannot possibly occur without significant social, political, and economic change.'

Such considerations are the source of great consternation for many. To begin with, at the moment, few people have any strong desire to live well past the age of 100 and beyond. But this is because our psychological software can't conceive of the idea of a 100-year-old who is still just as vital, agile, active, healthy and happy as a 40- or 50-year-old. It is just too much of an intellectual leap for us to alter our current mental image of a centenarian sufficiently. There is a good chance this will change over time, however, as there is an ever-increasing number of individuals in our society living to a much older age and living well in doing so.

Then there are the bigger philosophical issues as people contemplate the impact such an age 'revolution' could have on our global population and on climate change and environmental degradation too of course. Here, however, there is real cause for hope. If the evidence of history is anything to go by, there is a good chance that many of us living far longer, healthier lives could actually be incredibly positive for the human experience and for our planet, as counterintuitive as that may seem.

The last century or more has coincided with roughly a four-decade increase in average life expectancies in the developed world. In the same time frame things have become very materially better for most of humanity based on any more or less empirical assessment of the significant majority of the things that really matter. In the early nineteenth century it is estimated that roughly 90 per cent of the world's population lived in extreme poverty. Today that number is around 10 per cent. We have come a long way, and we have done so at the same time as a significant rise in life expectancy and population overall.

This is precisely the opposite outcome to that predicted by any number of the leading lights of the nineteenth century. I have already mentioned the doomsday predictions of Thomas Malthus, but he wasn't the only thinker of that age to forecast a horribly bleak future as a consequence of population growth. In his 1848 work, *Principles of Political Economy*, the great British philosopher, economist and politician John Stuart Mill argued that population growth was one of the major challenges facing society, and that it needed to be addressed through measures including birth control. Quite possibly taking her cue from folks like Mill, Queen Victoria fretted in a letter written to her daughter Princess Victoria in December 1858: 'I am most anxious about the population question. It is growing at a fearful rate and no one knows how to check it. I earnestly hope that measures may be taken in time to avert such a serious calamity to our country before we are overwhelmed by a multitude of paupers.' This position was perhaps unsurprising when you consider the abject squalor, disease and overcrowded misery that was Victorian London. Imagine her unquestionable surprise had she been able to visit present-day London with nearly 9 million inhabitants as against some way fewer than 3 million in her day. Present-day London is immeasurably greener, cleaner and more pleasant yet with more than three times the population that filled her with such dread.

No less a figure than Charles Darwin was another one who shared such views. In his 1871 work *The Descent of Man*, he worried that population growth would lead to competition, conflict and the spread of disease. On the other side of the Atlantic, Andrew Carnegie, the American industrialist and philanthropist, was similarly concerned. In his 1889 essay, 'The Gospel of Wealth', he feared that population growth would lead to serious social unrest and instability. Again, we can have some measure of sympathy for his stance. New York, Boston and other American cities at the time were just as awful, overcrowded, disease-ridden and squalid as London.

If you had suggested to any of these folk that more than a century after they had articulated their fears, the global population would indeed have increased, and more than sixfold, *but* that rates of illness, war, homicide and violence would have *declined* extraordinarily and that life expectancy and per capita wealth would have *increased* enormously, they would have thought you were entirely mad.

As counterintuitive as it may feel to us, our fears about another multi-decade expansion in life expectancy in future will likely be as ill-founded as theirs were because we, like they did, tend to heavily underestimate the power of human ingenuity and the role of exponentials. Of course, the counterargument is that we just haven't 'hit the buffers' yet. The fact that human population is now north of 8 billion *is* coming home to roost in terms of species extinction, habitat destruction, environmental degradation, climate change and competition for dwindling resources. Malthus, Mill, Queen Victoria, Darwin and Carnegie were right; they were just out by about one and a half centuries. The apocalyptic denouement they predicted is coming; it just hasn't arrived yet.

There is a good chance, however, that such fears may be overblown, if not just downright wrong. A key point here is that the more economically developed and well-educated people are, the lower their birth rate falls. And the older a population is, on average, the wealthier and better educated it is – another exponential function. When you put those two things together and combine them with the march of exponential scientific progress, there is a good chance that we may yet be able to deliver a future where all is not lost.

Population growth will slow and then quite possibly even go into reverse. This has already happened in most of the world's most developed nations if you control for the effect of immigration, including in Japan, Switzerland, Germany and much of Scandinavia. In the USA and UK the population is growing at a steadily slower rate – again, once you control for immigration.

At the same time, (bio)technology will continue to make our consumption patterns far more efficient. We will get vastly better at refashioning the periodic table and the natural world to provide food, energy, shelter and mobility without consuming and/or damaging nearly so much of it by revolutionizing agriculture, architecture, travel, packaging and power generation, for example.

We may move naturally towards a smaller population that lives longer, wealthier lives while causing significantly less damage to our environment overall – and this should happen iteratively and organically thanks to technological and cultural progress rather than violently or as a result of any kind of government edict.

It certainly hasn't been plain sailing over the last two centuries by any means, but the trajectory of humankind has most certainly been incredibly positive and the Malthusian pessimism of so many has proven to be widely off the mark for a long time. Let us hope that our fears about the next few decades turn out to be as unfounded as those articulated by the great thinkers of the nineteenth century.

In summary

In this part of the book we have looked at how the pharmaceutical and biotech industries have evolved over the last century or more and trotted through a number of the key technologies: from small molecules and onwards, via the discovery of DNA, towards large molecules, gene and cell therapies, and the promise of stem-cell treatments and gene editing using CRISPR. We have also considered the contribution made by advanced diagnostic technologies such as PCR and NGS and by artificial intelligence and machine learning, all of which are vital to the progress being made across the industry.

Over and above the great strides being made in therapeutics and diagnostics, the convergence of all of the above with new (bio)manufacturing techniques will also deliver powerful regenerative medicine in the near future. It may not be too long before we can bioengineer and 3D-print 'spare' organs for transplant surgery and before we have advanced tissue engineering capabilities. Fans of the Star Wars film franchise may remember Luke Skywalker losing his hand at the end of the second original film in that series, *The Empire Strikes Back* and having it replaced with a fully functional biomechanical version at the beginning of the next movie, *Return of the Jedi*. We may not be that many years away from this sort of technology crossing out of the realms of science fiction and being more or less widely available.

Given the pace of innovation, there will almost certainly be 'wild card' technologies and new approaches that emerge from left field which could take us all by surprise in terms of what they might deliver. One area which could be particularly interesting in this respect is the nascent field of nanotechnology.

There are already a number of on-market nanoparticle-based products being used in the clinic for diagnosis and drug delivery. Longer-term, nanoscale structures may be able to build functional tissues and organs, potentially revolutionizing the field of regenerative medicine and create implantable devices that could monitor and control biological processes in the body, such as insulin delivery for diabetes management, for example.

Similarly, our increasing understanding of the bacterial and viral worlds will very likely deliver new therapeutic treatments against a wide range of diseases and health conditions where dysbiosis has been implicated.

As we have also seen, all of this will have a role to play 'without us' as well as within us, much of which could be even more important for our future health and happiness as a species than improving our healthcare systems alone.

It will be exciting to see how all of these various emergent technologies develop in the years ahead.

Conclusion

Our future is biotech

'On what principle is it that with nothing but improvement behind us, we are to expect nothing but deterioration before us?'

Thomas Babbington Macaulay

The biotechnology industry has emerged as one of the most significant fields of science in recent years. Companies and research scientists working in the field are producing extraordinary technology and have created several trillion dollars' worth of real wealth in roughly four decades since the first biologic drugs were approved. The industry has the potential to revolutionize so much of what we do as a species and is making progress at an exponential rate, which makes it all the more exciting. From new medicines and medical devices embedding technology which would have been unthinkable only a few years ago, to how we grow, preserve and distribute our food, to technologies which could ensure that we have enough clean power and can do all the above while causing significantly less damage to the planet.

Biotech 'miracles'

The industry is already providing advanced treatments for previously incurable diseases. Biotech-based drugs are being used to treat various types of cancer, genetic disorders and autoimmune diseases like never before and may even be able to help us to 'treat' ageing. Even if biotech companies may increasingly be able to deliver what we might describe as 'miracle cures' in the near future, perhaps even more important will be the key role the industry has to play in moving us away from our current 'sick care' system to a kind of 'medicine 3.0' which will deliver far better health and longevity outcomes for millions of us, even billions of us in the fullness of time.

Earlier in the book we cantered through the importance of our microbiome and dysbiosis, nutrition, breathing, hormetic stress, including cold and heat exposure, sleep, and exercise and movement including strength-training over and above just aerobic or cardio-focused exercise alone. We also looked at the importance of the 'expectation effect' for all of the above – the idea that the more you believe in the merits of something, the more likely you are to derive benefit from it – and of growing evidence for the crucial importance of the mind–body connection. There is also a circular and self-reinforcing relationship between all of the above; if we eat and move better, we will sleep better. If we sleep better, we will experience lower levels of stress and be more inclined to exercise, eat well and have a positive outlook on life.

A successful move away from sick care towards the infinitely better healthcare and healthspan outcomes which will be achieved by 'medicine 3.0' requires many more of us to develop much better habits with respect to all of the above. Improvement here is all about 'little and often' behaviours throughout our entire lives. We need to spend far less time 'going on a diet' and far more time being on the right diet for the whole of our lives, and, crucially, one that is right for our own personal genome and biome, too.

It is the biotech industry which has given us the tools increasingly to understand the importance of all of the above in far more detail and it is the convergence of the tech and biotech industries which is spreading this information and giving us the apps, nutraceuticals, pharmaceuticals, medical devices, diagnostic tools and treatments which will help us embed such habits into our lives. When it comes to our diets, for example, historically there has been incredibly conflicting and confusing information about what the 'right' diet is. In future, this will be less of a problem as we come to understand our own personal and unique relationship with food and the composition of our own unique microbiome and even virome longer term.

Biotechnologies will deliver the diagnostic tools and wearable technologies which can finally optimize our diets and our microbiome, our exercise and our breathing and sleeping habits. Such technologies will soon become cheap enough to be available to many millions of us, not just the wealthiest early adopters in the developed world. David Sinclair makes the point that today most of us know significantly more about the status of our cars from their advanced dashboards than we do about ourselves. It won't be long before many more of us have access to a similarly sophisticated 'dashboard' when it comes to every aspect of our health and this will be transformational, enable many of us to live far healthier, happier and longer lives and save our healthcare systems billions, if not trillions, of dollars.

Perhaps even more important and exciting is the role the industry will play 'without' us, over and above 'within' us – the extent to which these technologies will improve the planet and our relationship with the natural world. Biotech companies will enable us to very significantly improve how we produce food. We will be able to develop crops that are resistant to pests and disease, thereby reducing the use of harmful pesticides and herbicides. We will be able to cultivate such crops under much harsher environmental conditions than in the past even, potentially, whilst achieving higher yields. All of this should be extremely positive for farmers in the developing world. Such technologies can help feed the world and improve food security. In the developed world, advanced hydroponics will enable us to grow crops locally and potentially far more efficiently than in the past once exponentials have played their part in terms of reducing cost and increasing efficiency.

Similarly, within one or two decades, biotech has the potential to revolutionize livestock farming and aquaculture. We will be able to enjoy lab-grown animal protein that is genetically identical to what we have obtained historically, but without needing to slaughter billions of unfortunate cows, pigs, sheep and chickens each year, or destroy the marine life in our oceans and seas. By the time we get there, this will seem perfectly normal to most of us. Just over a century ago the majority of people believed flying was entirely unnatural. Today billions of us fly. A few decades before that, people worried that trains travelling at more than 20 mph might cause madness or even death. Today millions of us travel at speeds approaching ten times that figure. We have a long track record as a species of adjusting to new technologies. Given how much better locally grown, bio-cultured meat and crops will be for all of us and for our planet we will likely do the same when it comes to such things, particularly when you consider the astonishing upside with respect to the environment and biodiversity that will result as we are able to rewild millions of acres of farmland across the globe.

The biotech industry will be instrumental in the development and continued exponential growth of renewable energy sources. New technologies will increase the efficiency with which we generate power from biomass, including algaculture (i.e., the use of algae we could grow in sea water in a tiny footprint of our oceans rather than on more valuable and scarce arable land), and from the sun with bio-coated PV cells, for example, and no doubt in myriad ways that we have yet to even contemplate. This will reduce pollution and our dependence on fossil fuels.

More generally when it comes to the environment, biotech companies will provide innovative solutions for some of the world's most pressing existing environmental problems. As we increasingly understand the role played by microorganisms, we will be able to degrade and detoxify pollutants in our environment. Powerful 'bioremediation' technologies will be used to clean up contaminated sites and restore damaged ecosystems. As we saw earlier in the book, microorganisms account for comfortably one thousand times the biomass of human beings. It stands to reason that improving our understanding of that world could pay enormous dividends across the piece.

Similarly, such technologies can and will revolutionize the packaging industry and many industrial processes. The biotech industry has already been developing biodegradable cellulose-based packaging products for some years. These should gradually replace single-use plastics in time. Such products can be made from renewable sources such as wood pulp or, increasingly, grown synthetically in a lab and can be broken down by natural processes, significantly reducing their environmental impact as compared to plastics. Cellulose-based packaging can be sturdy and versatile, making it suitable for a wide range of products and such materials are steadily becoming more cost-effective and widespread. The adoption of such products throughout the global supply chain will significantly reduce the amount of plastic waste that ends up in landfills and oceans.

Such products will also reduce oil consumption: the International Energy Agency (IEA) estimated that the production of plastics consumed approximately 359 million tonnes of oil in 2020. This accounted for about 5 per cent of global oil demand. Demand is still growing but it shouldn't be too many more years before 'natural' cellulose-based materials can slow and then reverse that trajectory.

Biotech companies will have a role to play in many other industrial applications beyond packaging. Biotech processes can improve the production of many industrial chemicals and reduce our reliance on petroleum-based products still further. The technology has the potential to reduce the carbon footprint and environmental damage caused by many industries. It will be exciting to watch how this develops, no doubt exponentially.

Overcoming challenges

Whilst the biotech industry has tremendous potential to improve, even revolutionize so much, there are also significant challenges to progress as we saw earlier in the book. One key problem is the ability of potentially world-changing companies and technologies to access sufficient capital, particularly outside of the US, for all the historical and cultural reasons we explored.

Biotech companies require many millions, often even billions of capital to do what they do. Smaller companies in particular can struggle to secure sufficient funding for their research, and even larger companies confront the high costs associated with clinical trials, regulatory approvals, manufacturing and sales and marketing activities.

None of this is helped by structural inefficiencies in how such companies are funded, even in modern developed world places such as the UK, Europe, Australasia and beyond. To a great extent we are fighting with one hand tied behind our back in terms of the progress we could be making.

Related to that lack of capital is a real public perception problem so often faced by the industry. Too often biotech research is met with scepticism and fear from the general public. There is a powerful notion of 'bad pharma' which, all too frequently, can cast a shadow over so many companies doing wonderful things or who would be able to do wonderful things with a bit more capital and media coverage.

When it comes to the very cutting edge of what the science can do whether that be in gene editing, GM crops or a wish to see us all live long into our second centuries, the industry can also too often be seen as wishing to 'play God'. This arguably limits capital flows still further and, in modern democracies, can make governments less inclined to support the industry than they otherwise might be, particularly if our politicians share the same misgivings as the general public.

As I hope I've shown throughout the book, many of these fears are overblown or even just plain wrong. All too often concerns about the industry arise because people know little or nothing about it. As David Sinclair has put it: 'Being wrong has never stopped conventional wisdom from negatively impacting public policy.' Or as Jennifer Doudna has put it: 'society cannot make decisions about technologies it doesn't understand, and certainly not about those it knows nothing about.'

Yet it seems likely that these challenges will be overcome, if only thanks to the inexorable march of exponential technological progress. The very nature of exponentials is that, eventually, they tend to be more or less irresistible.

Robert Metcalfe is an American engineer and former professor and the winner of several of the technology industry's most prestigious awards, including the US National Medal of Technology and Innovation and the ACM Turing Award. He was a key contributor to the development of the internet and co-invented Ethernet. In 1980 he proposed what was soon dubbed 'Metcalfe's law' (Figure 12.1), one of the key ideas which has underpinned the growth in telecommunications, the internet and the network economics which are a key explanation for the value created by the technology industry in the last few decades.

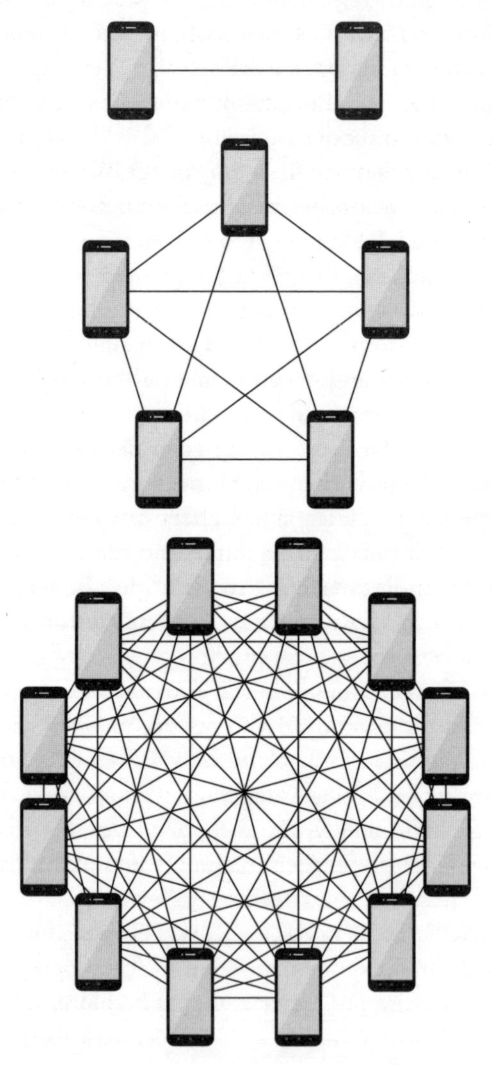

FIGURE 12.1 Metcalfe's law: two mobiles can make only one connection, five can make ten connections, and 12 can make 66 connections.
Adapted from Wikipedia.com, under Wikipedia Commons

The law states that the value of a network develops *exponentially* because it is *proportional* to the *square* of the number of connected users, devices or 'nodes' in a given network. The progression and value creation in any network are therefore geometric, not linear.

The law was originally applied to fax machines, telephones, computers and servers but is certainly relevant to the trajectory of the biotech industry. First, and somewhat prosaically, because the industry uses those telephones, computers and servers to share large datasets as we have already highlighted but, more generally, because the total number of biotech scientists in the world is a network like any other.

The more biotech scientists there are in the world, in aggregate, the greater the potential for collaboration and innovation and the development of new ideas, therapies and technologies. Thanks to Metcalfe's law, as the total number of biotech scientists, R&D spend and patent registration grows, the impact of that growth on scientific output will be exponential rather than linear.

In the context of a global population of more than 8 billion of us, there are still relatively few biotech scientists. Related to this, it is also the case that a great deal of what we have covered in the last many pages has all happened relatively recently. The exponentials are only just getting started. CRISPR, next-generation sequencing, effective AI and ML have all only been with us for a few years, for example, and the processing power and communications technology needed to make the most of those technologies have arrived only in the very recent past. Across the board we are actually only in the very early days of the proverbial exponential 'lily pond'.

It also seems likely that many of the funding challenges faced by biotech companies outside of the US will resolve in time – initially because enlightened global investors are increasingly taking note of the valuation opportunity outside of the USA. While the market may have been in disequilibrium for the last many years, in the medium to long term, eventually the 'efficient market hypothesis' *will* apply and global capital will flow towards undervalued assets.

This process will be accelerated as a crop of companies in places like the UK, Europe and Australia deliver commercially and become more attractive to generalist investors and overseas investors alike as a natural function of that commercial delivery. Earlier in the book we looked at how the UK has struggled to develop the ecosystem required to nurture and grow the biotech industry as well as it might. In time, however, it is more than likely that the UK and other countries will get there, if only as a function of the combination of the exponential development of the science, global capital flows and commercial delivery by a handful of corporate success stories and, dare we hope, the extent to which a few of those become front-page news. As Welsh scientist, entrepreneur and venture capitalist Sir Chris Evans has put it, 'the UK only needs one Genentech'. Given the quality of British science, we might hope that this will happen in the fullness of time and elsewhere in Europe and Asia too.

Overcoming our fear of biotech

Exponential progress in these technologies is also not something we should fear. As we saw earlier in the book in the section on CRISPR, researchers and scientists working on these technologies are showing themselves to be good actors in the development of cutting-edge technologies which raise moral and ethical concerns. There are also plenty of checks and balances from a wide range of stakeholders and institutions including government regulators of course.

It is worth stressing that we have been 'genetically modifying' crops and livestock for thousands of years. A little-known example of this fact is that the orange colour that we now commonly associate with carrots is the result of selective breeding by Dutch horticulturalists. Throughout the seventeenth century Dutch farmers selectively bred yellow and white varieties to produce an orange-coloured carrot. Some claim that this was in honour of the Dutch royal family, the House of Orange-Nassau, although whether this is true or not is the subject of some debate. What is true, however, is that, over time, the orange carrot became the most popular and widely cultivated variety, and it eventually became the norm in many parts of the world. Similarly, large red tomatoes have come about only as a result of thousands of years of focused cultivation. None of us have a problem with orange carrots or large red tomatoes, at least not in my experience, even though they are both 'GM crops' in the literal sense.

Selective breeding is the reason we have hornless cattle which produce higher volumes of milk and more meat than their ancestors did. As Jennifer Doudna points out, this is why we can find a four-pound Chihuahua playing alongside a two-hundred-pound Great Dane, despite both being part of the same species. The only difference between today's techniques and those employed for the last several centuries is that they are now far more accurate, effective and fast. Today, biotech allied to big data and computational power can do in days or hours what previously took generations. As

Doudna elaborates: 'What is breeding but another tool of genetic manipulation, like CRISPR only less predictable and efficient?'

If anything, we need fewer restrictions on the progress of this science, not more. All too often objections are based on bad science and excessive risk aversion and pessimism which, for reasons of deep-seated human psychology which we looked at earlier in the book, can be more powerful for our collective consciousness than the happier reality. Most of our worst fears around technologies such as GM crops or gene editing will no doubt prove to be unfounded.

We have a number of examples of this dynamic throughout human history, perhaps the most recent of which is our experience since the Manhattan Project ushered in the nuclear age some 80 years ago. Aside from a handful of admittedly pretty awful events such as occurred at Three Mile Island, Chernobyl and Fukushima, our worst fears about the advent of the nuclear age have certainly never been realized.

While I am confident that the industry will overcome the challenges I've presented in this book and deliver so much for us, it could certainly do with a little help from a PR perspective, if we are to move as fast as we could at least. We need a far larger percentage of the population to be aware of the exciting progress being made, not just a small elite of the most techno-literate first adopters in places like Silicon Valley. And we need more people to be spreading the positive aspects of these technologies and far fewer to endlessly telegraph their concerns about 'bad pharma'.

It is all too easy to find examples of bad actors in every walk of life, but in doing so I think we risk focusing on the minority of things that are negative and missing vastly more that is positive, uplifting and to be celebrated. We must guard against throwing the proverbial baby out with the bathwater (as it were). The more of us there are focused on and talking about those positives, the better.

As we saw earlier in the book more or less empirically, the more people understand about things, the more likely they are to be positively inclined towards them. I hope that this book may have made a contribution to that reality, even if only in a small way. One of my all-time favourite quotes is John F. Kennedy's famous exhortation: 'Ask not what your country can do for you – ask what you can do for your country.' As we perhaps move into an era where nation states are less important than in the past and where more of us like to think of ourselves as global citizens, might we reframe this in terms of asking what each of us can do for humanity and for progress overall?

Investing in biotech

To that end, over and above being more positively inclined towards the biotech industry and excited about what it may deliver in the near future and, dare I hope, sharing that position far and wide, the other thing you might consider as this book draws to a close is to invest in it. As American investor, author and entrepreneur Jim O'Shaughnessy has put it: 'Pessimists sound smart, optimists make money…'

My first book suggested that it might be a good idea to 'own the world'. The most fundamental idea underpinning that suggestion was that throughout history, human progress has been the most important investment theme of them all. My suggestion to own the world was all about doing our best to have investment exposure to that reality.

Given the value that the biotech industry will likely create in the next few decades, it may be a good idea to have some exposure to it as an investor. This can be achieved more or less easily by simply owning a large stock market index such as the S&P 500 in the USA or the MSCI World or All-Country World global indices. As the biotech industry grows and develops it will make a contribution to the aggregate level of those indices, *directly* as biotech companies grow and *indirectly* as the technologies help deliver value across many other key industry sectors such as agriculture, chemicals, technology and energy just as has occurred with the tech industry over the last many decades.

For most of us, 'owning the world' this way will give us at least some exposure as the industry develops. Those who are older, wealthier or more financially sophisticated might consider some more 'pure' exposure as long as they're mindful of the potential risks.

There is an astonishing amount to be excited about. We can navigate the challenges successfully to a series of wonderful outcomes.

Our future truly is biotech and given the direction of travel there is a good chance that it is more than likely to be a bright one for us as a species and for the natural world, too…

Please help spread the word!

Key resources for health, happiness and longevity

As I said in the introduction, one of the 'tangible benefits' I want you, the reader, to get out of this book is the ability to improve your physical and mental health. To that end, below I have attempted to draw together a summary of some of the key ideas covered and provide links to additional resources.

In Part 2, we looked at the rise of many 'diseases of modernity' or 'modern plagues', including numerous debilitating allergies and autoimmune diseases, IBD, IBS, obesity, diabetes, depression and other mental illnesses. In Chapter 11, we also looked at the subject of longevity and at the idea that we may be able to slow down the aging process. The main benefit of seeking to do this is to reduce the risk of all sorts of diseases and health conditions throughout your life as well as potentially extending your 'healthspan'.

Whether you or a loved one are suffering from a chronic health condition or you simply want to optimize your health, I hope that the ideas and resources below will help.

Addressing dysbiosis

In Chapter 5, we looked at how the microbial world has a key role to play in our health and mental health and, in particular, at the extent to which *dysbiosis* – a compromised or imbalanced microbiome – is being implicated in disease after disease and numerous mental health conditions. One of the most fundamental building blocks for vibrant health then seems likely to be to optimize our microbiome and repair dysbiosis.

If you wish to find out more about this subject, I can highly recommend the following books:

- *10% Human: How Your Body's Microbes Hold the Key to Health and Happiness* by Alanna Collen

- *Gut: The Inside Story of Our Body's Most Under-rated Organ* by Giulia Enders
- *Missing Microbes: How Killing Bacteria Creates Modern Plagues* by Martin Blaser

In terms of tangible steps to seek to combat dysbiosis, in Chapter 6 we looked at the merits of fermented foods. These include: kefir, sauerkraut, natto beans and miso, kombucha, kimchi and a great deal else besides.

I also highlighted my personal experience of a probiotic product called Symprove which I credit with combatting my own IBD. You can find out more at their website: www.symprove.com

Physical health – diet and supplementation

In Chapter 6 we looked at the key roles played by diet, breathing, hormetic stress, sleep and exercise and movement.

With respect to diet and nutrition, one of *the* key ideas here is that there really is no such thing as a one-size-fits-all diet. The 'right' diet for you is a complex function of your own personal genome and microbiome as we have seen. It is important to try a wide variety of diets to establish what works best for you and stay clear of dogmatic ideas that insist 'vegan' or 'carnivore' diets are necessarily 'best'.

If you are seriously ill, confronting a troublesome health condition or just want to optimize how you feel, it could be worth using one of the genetic testing companies available. Examples of such companies include 23andMe (www.23andme.com) and Zoe (www.zoe.com). There are plenty of others depending on where you are based. You might also consider the merits of a nutritionist if you have been dealing with a particularly intractable health problem.

This having been said, another crucial point is that we should certainly be more focused on what TO eat rather than what NOT to eat. Increasingly it seems that nutrient *inadequacy* over many years (rather than deficiency) is a key causal factor for so many chronic diseases and health conditions as we saw in Chapter 6 when I referenced the work of Dr Rhonda Patrick.

To that end, I believe there is plenty to recommend the idea of some key supplementation to attempt to cover as many nutritional bases as possible. I personally use:

- Athletic Greens / AG1 (www.drinkag1.com/en-uk) – endorsed by athletes and clinicians alike this is a powerful cocktail of vitamins, minerals, bacterial cultures and botanicals.
- Vitamin D – numerous studies have found that vitamin D deficiency is a huge problem in much of the world. For this reason I take it daily.

A number of studies have shown that the two main omega-3 fatty acids (EPA and DPA) may also have a number of material long-term health benefits. Rhonda Patrick's website has a solid article on the subject (https://www.foundmyfitness.com/topics/omega-3) and it may be worth supplementing with a good omega-3 product if you suspect you don't get enough in your diet.

We also looked at the idea of 'intermittent fasting (IF) in Chapter 6. The idea is to trigger 'autophagy' – the breakdown and recycling of damaged cells and cellular components – which numerous studies have found is intimately linked to health. There are various approaches to IF with the main two being:

1. A *daily* approach – where you ensure that you don't eat for many hours of each day. The easiest way to do this is to eat dinner relatively early and breakfast relatively late so that you 'fast' overnight. Some advocates refer to this as 16:8 – in that you fast for 16 hours and eat all your calories within 8 hours of any given 24-hour period.
2. The 5:2 diet – where you eat normally five days a week then have only one small meal on two *non-consecutive* days of the week.

My preference is the daily approach where I eat dinner at about 6pm and breakfast at about 8.30am. This is more like 14 hours 'off' but pretty easy to do and which the current science seems to suggest will have at least some benefit.

Physical health – breathing, hormetic stress and sleep.

In terms of breathing and oxygen (the 'forgotten nutrient'), hormetic stress through cold and heat exposure, and getting enough sleep, the following books are well worth a look on these subjects:

- *Breath: The New Science of a Lost Art* by James Nestor

- *The Wim Hof Method: Activate Your Potential, Transcend Your Limits* by Wim Hof
- *Why We Sleep: The New Science of Sleep and Dreams* by Matthew Walker

In terms of easy steps you might take to implement some of these ideas:

- I do Wim Hof's breathing exercises once or twice a day – early in the morning and after lunch (https://www.wimhofmethod.com/breathing-exercises).
- For the last 1–2 minutes of my shower each morning I gradually switch it to as cold as I can bear. This feels great (nearly) every morning by the way … afterwards at least.
- I'm also fortunate that there is a sauna and steam room at my gym but a very hot bath or three each week should help if you don't have access to a sauna or steam room.
- I use the 'Sleep Cycle' app every night (https://www.sleepcycle.com/) – there are plenty of others.

Physical health – exercise and movement

In Chapter 6 we looked at the importance of VO2 max, physical strength, movement and flexibility for health and longevity in particular. Two great books here are:

- *The 4-Hour Body* by Tim Ferriss
- *Built to Move: The 10 Essential Habits that will Help You Live a Longer, Healthier Life* by Kelly and Juliet Starrett

Tim Ferriss's idea of 'minimum effective dose' is particularly helpful – in that you don't need to do that much to make a big difference. Over a year, even only one resistance weights session in the gym per week will still be 52 times more than none at all, and far easier to achieve that than trying to go every day.

I've been doing 10–15 press ups and a simple stretch routine most mornings for several years now. I also go for a walk in the morning pretty much every day of the year and use the Under Armour 'Map My Run' app' to log as many steps as possible (https://www.mapmyrun.com/). There are plenty of other more sophisticated options nowadays.

Mental health

As we saw in Chapter 7, mental health and physical health are intimately related and our expectations can have a tangible effect on outcomes too. We also looked again at the merits of meditation and at the crucial importance of habit and 'little and often'. Key books you might consider with respect to all of the above include:

- *The Expectation Effect: How Your Mindset Can Transform Your Life* by David Robson
- *Atomic Habits: An Easy and Proven Way to Build Good Habits and Break Bad Ones* by James Clear
- *10% Happier: How I Tamed the Voice in My Head, Reduced Stress Without Losing My Edge, and Found Self-Help That Actually Works – A True Story* by Dan Harris
- *Tools of Titans: The Tactics, Routines, and Habits of Billionaires, Icons, and World-Class Performers* by Tim Ferriss
- *Tribe of Mentors: Short Life Advice from the Best in the World* also by Tim Ferriss

Dan Harris's book is particularly good if you're more than a little cynical about meditation (as he was originally – a key theme in the book).

Nowadays there are plenty of apps which can be very helpful here. For daily meditation I love the 'Calm' app (www.calm.com) and have also used 'Headspace' in the past (www.headspace.com).

For what it is worth, over and above meditation, there is also growing evidence for the health and mental health benefits of a gratitude practice. Gratitude combats depression and anxiety and has been shown to reduce stress, improve cardiac health and sleep.

To that end, I'm also a huge fan of the '5 Minute Journal' app from a company called Intelligent Change which can be downloaded on iOS and Android phones (https://www.intelligentchange.com/pages/our-story).

Longevity specific resources

Everything we have looked at above should be helpful for longevity and healthspan. If you would like to learn more on the subject, I would recommend these two books:

- *Lifespan: Why We Age – and Why We Don't Have To* by David Sinclair
- *Outlive: The Science and Art of Longevity* by Dr Peter Attia

If you spend any time looking into longevity, you will see that there are a number of supplements which some of the leading lights in the space are known for taking. My reading is that there is still a fair bit of controversy around such things. As such, I don't take any of them personally and have omitted any reference to the specifics here.

I hope this brief summary of some of the key concepts from the book has given you a practical toolkit and reasonably concise 'A to B' of resources to help you implement some of the core ideas.

References

Chapter 1

Azhar, Azeem, *Exponential: Order and Chaos in an Age of Accelerating Technology*. London: Penguin, 2022.

Black Rock – Larry Fink's CEO letter: https://www.blackrock.com/corporate/investor-relations/2020-larry-fink-ceo-letter

Yale Climate Connections https://yaleclimateconnections.org/2021/01/investors-flee-big-oil-as-portfolios-get-drilled/

Jurvetson, S. 'Transcending Moore's law to forge the future', *Core Magazine*, 2015, 36–9.

Linklaters, Insights Publication, July 2020

Senior, M. 'Innovators take cover as market bubble bursts', *Nature Biotechnology* 40.4 (2020): 450–7.

Unilever press release: https://www.unilever.com/news/press-and-media/press-releases/2020/unilever-sets-out-new-actions-to-fight-climate-change-and-protect-and-regenerate-nature-to-preserve-resources-for-future-generations/

Chapter 2

Carey, N. *Hacking the Code of Life: How Gene Editing Will Rewrite Our Futures*. London: Icon Books, 2019.

Dormandy, A., quoted in P. Foster and D. Thomas, 'The UK's dream of becoming a "science superpower"', *Financial Times*, 5 January 2023.

Duncan, E. 'Don't blow Britain's great life sciences chance', *The Times*, 26 January 2023.

Gawande, A. *The Checklist Manifesto: How to Get Things Right*. New York: Metropolitan Books, 2009.

Gawande, A. 'The future of medicine', *Reith Lectures 2014*, BBC Radio 4, https://www.bbc.co.uk/programmes/b04bsgqn

Herbert, A. S. 'Rational decision-making in business organizations'. Stockholm: The Nobel Foundation, 1979.

Nave, C., quoted in S. Rose, 'Brandon Capital raises $200m from four industry funds for medical VC', *Financial Review*, 19 April 2015, https://www.afr.com/markets/equity-markets/brandon-capital-raises-200m-from-four-industry-funds-for-medical-vc-20150417-1mn3vl

Nobel Prize press release – Richard H. Thaler: https://www.nobelprize.org/prizes/economic-sciences/2017/press-release/

Royal College of Surgeons. *From Innovation to Adoption: Successfully Spreading Surgical Innovation*, report, 2014, https://www.rcseng.ac.uk/library-and-publications/rcs-publications/docs/from-innovation-to-adoption/

Sanghera, G., quoted in P. Foster and D. Thomas, 'The UK's dream of becoming a "science superpower"', *Financial Times*, 5 January 2023.

Chapter 3

Ambrose, J. 'Renewable electricity overtakes fossil fuels in UK for first time', *The Guardian*, 14 October 2019.

Azhar, Azeem, *Exponential: Order and Chaos in an Age of Accelerating Technology*. London: Penguin, 2022.

Becher, J. 'Lily pads and exponential thinking', *Manage by Walking Around* blog, 31 January 2016, https://jonathanbecher.com/2016/01/31/lily-pads-and-exponential-thinking/

Botsman, R., and Rogers, R. *What's Mine Is Yours: How Collaborative Consumption is Changing the Way We Live.* London: Collins, 2011.

Bouwer, J., Saxon, S., and Wittkamp, N. 'Back to the future? Airline sector poised for change post-COVID-19', mckinsey.com, https://www.mckinsey.com/industries/travel-logistics-and-infrastructure/our-insights/back-to-the-future-airline-sector-poised-for-change-post-covid-19

Cancer.org: https://www.cancer.org/latest-news/facts-and-figures-2021.html

Cuthbertson, A. 'World record achieved for solar power "miracle material"', *The Independent*, 13 April 2022.

Diamandis, P., and Kotler, S. *Abundance: The Future Is Better Than You Think.* New York: Free Press, 2012.

Future Crunch: https://futurecrunch.com/about/

Mellon, J. *Moo's Law: An Investor's Guide to the New Agrarian Revolution.* Sudbury: Fruitful Publications, 2020.

Ocasio-Cortez, A., interview with interview with writer Ta-Nehisi Coates at the MLK Now event in New York City, 21 January 2019, https://www.youtube.com/watch?v=q3-QvoIfpxc

Oxford Science Enterprises, https://www.oxfordscienceenterprises.com/

Thunberg, G., Speech to the Houses of Parliament, given in full in *The Guardian*, 23 April 2019.

Wigley, T. quoted in M. Schellenberger, 'Why apocalyptic claims about climate change are wrong', *Forbes*, https://www.forbes.com/sites/michaelshellenberger/2019/11/25/why-everything-they-say-about-climate-change-is-wrong/

Wilson, E. O. *Consilience: The Unity of Knowledge*. New York: Alfred A. Knopf, 1999.

Chapter 4

Association of the British Pharmaceutical Industry: https://www.abpi.org.uk/media/news/2021/july/new-research-reveals-attitudes-to-pharmaceutical-industry/

Brower, N. 'Mind–body research moves towards the mainstream', *EMBO Reports* 7.4 (2006): 358–61.

Center for Strategic and International Studies (CSIS): https://www.csis.org/analysis/need-leapfrog-strategy

Centers for Disease Control and Prevention: https://www.cdc.gov/vaccinesafety/ensuringsafety/monitoring/vaers/index.html

Davies, S. C. *The Drugs Don't Work: A Global Threat*. London: Penguin, 2013.

Dyvik, Einar H. Total global spending on research and development, 1996–2022. Statista: https://www.statista.com/statistics/1105959/total-research-and-developmentspending-worldwide-ppp-usd/

Gallup: https://news.gallup.com/poll/266060/big-pharma-sinks-bottom-industry-rankings.aspx

Goldacre, B. *Bad Pharma: How Medicine is Broken, and How We Can Fix It*. London: Fourth Estate, 2013.

Grayling: https://grayling.com/news-and-views/grayling-research-reveals-increasing-goodwill-towards-pharma-but-can-it-last/

Harvard Medical School. 'The power of the placebo effect', *Harvard Health Publishing*, 13 December 2021, https://www.health.harvard.edu/mental-health/the-power-of-the-placebo-effect

Harvard School of Public health: https://www.hsph.harvard.edu/news/hsph-in-the-news/poll-shows-americans-are-fed-up-with-pharmaceutical-industry/

Kone, D. 'Advancing sanitation: 10 years of reinventing the toilet for the future', 28 July 2021, https://www.gatesfoundation.org/ideas/articles/sanitation-reinvent-toilet

National Institutes of Health (NIH): https://www.ncbi.nlm.nih.gov/pmc/articles/PMC7120529/

Pollan, M. *How to Change Your Mind: What the New Science of Psychedelics Teaches Us about Consciousness, Dying, Addiction, Depression, and Transcendence.* New York: Penguin, 2019.

Ridley, M. *The Rational Optimist: How Prosperity Evolves.* London: Fourth Estate, 2010.

Rosling, H., and Rosling, O. *Factfulness: Ten Reasons We're Wrong about the World – and Why Things are Better Than You Think.* London: Sceptre, 2018

UCI School of Pharmacy & Pharmaceutical Sciences: https://pharmsci.uci.edu/programs__trashed/a-short-history-of-drug-discovery/

Chapter 5

AMR Action Fund: https://www.jdsupra.com/legalnews/uk-life-sciences-and-healthcare-95161/

Collen, A. *10% Human: How Your Body's Microbes Hold the Key to Health and Happiness.* Glasgow: William Collins, 2016.

Davies, S. C. *The Drugs Don't Work: A Global Threat.* London: Penguin, 2013.

Destiny Pharma: https://www.destinypharma.com/xf-platform/

Gyungcheon Kim, 'Delayed establishment of gut microbiota in infants delivered by caesarean section', *Frontiers in Microbiology* 10 (September 2020).

Okada, H., et al., 'The "hygiene hypothesis" for autoimmune and allergic diseases: an update', *Clinical and & Experimental Immunology* 160:1 (2010).

Mayo Clinic: https://www.mayoclinic.org/diseases-conditions/childhood-asthma/expert-answers/hygiene-hypothesis/faq-20058102

ScienceDaily, 'Genetic census of the human microbiome', *ScienceDaily.com*, 14 August 2019, https://www.sciencedaily.com/releases/2019/08/190814113936.htm

Schuijs, M. J., et al. 'Farm dust and endotoxin protect against allergy through A20 induction in lung epithelial cells', *Science* 349.6252 (2015): 1106–10.

World Health Organization: https://www.who.int/news-room/fact-sheets/detail/antimicrobial-resistance

Chapter 6

Attia, P., in interview with Tim Ferriss, https://tim.blog/2023/03/17/peter-attia-outlive-transcript/

bd-capital: press release, https://bd-cap.com/symprove

Blackburn, E., and Epel, E. *The Telomere Effect: A Revolutionary Approach to Living Younger, Healthier, Longer*. London: Orion Spring, 2017.

Buerk, M., *Radio Times*, reported in https://news.sky.com/story/michael-buerk-let-obese-people-die-early-to-save-nhs-money-11778620

Center for Nutrition Studies, https://nutritionstudies.org/the-china-study/

Cheng, W.Y. 'The role of gut microbiota in cancer treatment: friend or foe?', *Gut* 69 (2020): 1867–76.

Collen, A. *10% Human: How Your Body's Microbes Hold the Key to Health and Happiness*. Glasgow: William Collins, 2016.

Dillon, S. 'We learn nothing about nutrition, claim medical students', BBC News, https://www.bbc.co.uk/news/health-43504125

Fattorusso A, Di Genova L, Dell'Isola GB, Mencaroni E, Esposito S. 'Autism Spectrum Disorders and the Gut Microbiota', Nutrients. 2019 Feb 28;11(3):521

Ferriss, T. *Tools of Titans: The Tactics, Routines, and Habits of Billionaires, Icons, and World-Class Performers*. London: Vermilion, 2016.

Greger, M. *How Not to Die: Discover the Foods Scientifically Proven to Prevent and Reverse Disease*. New York: Macmillan, 2016.

Healthline: https://www.healthline.com/health/food-nutrition/nutrigenomics-might-be-the-future-of-how-you-eat

Hof, W. *The Wim Hof Method: Activate Your Potential, Transcend Your Limits*. London: Rider, 2020.

Ivanova, Y. M., and Blondin, D. P. 'Examining the benefits of cold exposure as a therapeutic strategy for obesity and type 2 diabetes', *Journal of Applied Physiology* 130.5 (2021): 1448–59

Janus, J., 'Nutritional genomics', https://www.phgfoundation.org/explainer/nutritional-genomics-explainer-july-2021

Liu, B.-L. et al., 'Gut microbiota in obesity', *World Journal of Gastroenterology* 27.25 (2021): 3837–50.

Nestor, J. *Breath: The New Science of a Lost Art*. London: Riverhead Books, 2020.

Nutritank: https://nutritank.com/

Patrick, R., interview with T. Ferriss, https://tim.blog/wp-content/uploads/2018/09/12-rhonda-patrick.pdf

Schmidt, E.: https://tim.blog/2021/10/25/eric-schmidt-ai/

ScienceDaily, 'The diversity of rural African populations extends to their microbiomes', *ScienceDaily.com*, 22 January 2019, https://www.sciencedaily.com/releases/2019/01/190122084404.htm

Stanford School of Medicine: https://med.stanford.edu/news/all-news/2021/07/fermented-food-diet-increases-microbiome-diversity-lowers-inflammation

Starrett, J., and Starrett, K. *Built to Move: The 10 Essential Habits That Will Help You Live a Longer, Healthier Life.* London: Orion Spring, 2023.

Symprove: https://www.symprove.com/pages/learn-science

University of Oxford's Department of Psychiatry. 'Genes and mental illness', https://www.psych.ox.ac.uk/news/genes-and-mental-illness, citing M. Taquet et al., 'A structural brain network of genetic vulnerability to psychiatric illness', *Molecular Psychiatry* 26 (2021), 2089–100.

Valcarce-Torrente, M., et al., 'Influence of fitness apps on sports habits, satisfaction, and intentions to stay in fitness center users: an experimental study', *International journal of Environmental Research and Public Health* 18.19 (2021): 10393.,

Walker, M. *Why We Sleep: The New Science of Sleep and Dreams*. London: Penguin, 2018.

Willcox, D. C., 'The Okinawan diet: health implications of a low-calorie, nutrient-dense, antioxidant-rich dietary pattern low in glycemic load', *Journal of the American College of Nutrition* 28, supl. issue 4 (2009): 500S–516S.

Chapter 7

Blaser, M., *Missing Microbes: How the Overuse of Antibiotics is Fueling Our Modern Plagues*. New York: Henry Holt & Co., 2014.

Clear, J. *Atomic Habits*. London and New York: Penguin, 2018.

Robson, D. *The Expectation Effect: How Your Mindset Can Transform Your Life*. Edinburgh: Canongate, 2022.

Shapiro, D. *Your Body Speaks Your Mind: Understand the Link between Your Emotions and Your Illness.* London: Piatkus, 2007.

Chapter 8

Diamond, J. *Guns, Germs, and Steel: A Short History of Everybody for the Last 13,000 Years.* New York: Vintage, 1998.

Garde, D., and Saltzman, J. 'The story of mRNA: How a once-dismissed idea became a leading technology in the Covid vaccine race', *STAT*, 10 November 2020, https://www.statnews.com/2020/11/10/the-story-of-mrna-how-a-once-dismissed-idea-became-a-leading-technology-in-the-covid-vaccine-race/

Jenner Institute: https://www.jenner.ac.uk/about/edward-jenner

Lanese, N. '2 scientists win $3 million "Breakthrough Prize" for mRNA tech behind COVID-19 vaccines', *Livesciemce.com*, 9 September 2021, https://www.livescience.com/breakthrough-prize-winners-mrna-vaccines.html

Mackenzie, R. 'DNA vs. RNA – 5 key differences and comparison', *Technology Networks*, 18 December 2020, https://www.technologynetworks.com/genomics/articles/what-are-the-key-differences-between-dna-and-rna-296719

Mellon, J. *Cracking the Code: Understand and Profit from the Biotech Revolution That Will Transform Our Lives and Generate Fortunes*. Hoboken, NJ: Wiley, 2012.

The Nobel Prize in Physiology or Medicine 1945, http://www.nobelprize.org/nobel_prizes/medicine/laureates/1945/

Riedel, S. 'Edward Jenner and the history of smallpox and vaccination', *Baylor University Medical Center Proceedings* 18.1 (2005).

US National Library of Medicine: https://profiles.nlm.nih.gov/spotlight/sc/feature/doublehelix

Chapter 9

Australian Trade and Investment Commission (Austrade): https://www.austrade.gov.au/news/success-stories/adalta-innovative-shark-inspired-antibodies-to-revolutionise-treatments-for-serious-disease

Baltimore, D., et al. 'A prudent path forward for genomic engineering and germline gene modification', *Science* 348.6230 (2015): 36–8.

Caldwell, K. J., Gottschalk, S., and Talleur, A.C. 'Allogeneic CAR cell therapy – more than a pipe dream, *Frontiers in Immunology* 11 (2021), https://www.frontiersin.org/articles/10.3389/fimmu.2020.618427/full

Cell Guidance Systems, blog post, 3 July 2020, https://www.cellgs.com/blog/significant-challenges-remain-for-ipsc-based-therapeutics.html

Delea, T. 'Cost-effectiveness of blinatumomab versus salvage chemotherapy in relapsed or refractory Philadelphia-chromosome-negative B-precursor acute lymphoblastic leukemia from a US payer perspective', *Journal of Medical Economics* 20.9 (2017): 911–22.

Doudna, J., and Sternberg, S. *A Crack in Creation: The New Power to Control Evolution*. New York: Houghton Mifflin Harcourt, 2017.

Exscientia: https://www.exscientia.ai

Food and Drug Administration: https://www.fda.gov/vaccines-blood-biologics/cellular-gene-therapy-products/what-gene-therapy

Forbes, P. '*A Crack in Creation* review – Jennifer Doudna, Crispr and a great scientific breakthrough', *The Guardian*, 17 June 2017, https://www.theguardian.com/books/2017/jun/17/a-crack-in-creation-by-jennifer-doudna-and-samuel-sternberg-review

Formela, J.-F., and Stanford, J., 'To spend less on health care, invest more in medicines', *STAT*, 5 April 2022, https://www.statnews.com/2022/04/05/we-should-spend-more-on-prescription-drugs-not-less/

Gallagher, J. 'Sickle cell: "The revolutionary gene-editing treatment that gave me new life"', *BBC News*, 20 February 2020, https://www.bbc.co.uk/news/health-60348497

Gardner, S., quoted in https://precisionlife.com/news-and-events/indx-paper-reveals-potential-to-systematically-reposition-hundreds-of-patented-drugs-into-new-indications-to-address-unmet-medical-needs

Grupp, S., quoted in https://www.novartis.com/news/media-releases/novartis-five-year-kymriah-data-show-durable-remission-and-long-term-survival-maintained-children-and-young-adults-advanced-b-cell-all

Hopkins, A., quoted in N. Savage, 'Tapping into the drug discovery potential of AI', *Nature*, 27 May 2021, https://www.nature.com/articles/d43747-021-00045-7

Isaacson, W. *The Code Breaker: Jennifer Doudna, Gene Editing and the Future of the Human Race*. New York: Simon & Schuster, 2022.

'NIH Director's Blog': https://www.nih.gov/crispr-based-anti-viral-therapy-could-one-day-foil-flu-covid-19

Oxford Biomedica: https://oxb.com/oxford-biomedica-announces-rd-collaboration-with-microsoft-to-improve-gene-and-cell-therapy-manufacturing-using-the-intelligent-cloud-and-machine-learning/

Phillips, A., quoted in https://oxb.com/oxford-biomedica-announces-rd-collaboration-with-microsoft-to-improve-gene-and-cell-therapy-manufacturing-using-the-intelligent-cloud-and-machine-learning/

Sufian, S., and Garland-Thomson, G. 'The Dark Side of CRISPR', *Scientific American*, 16 February 2021, https://www.scientificamerican.com/article/the-dark-side-of-crispr/

Chapter 10

Mellon, J. *Cracking the Code: Understand and Profit from the Biotech Revolution That Will Transform Our Lives and Generate Fortunes.* Hoboken, NJ: Wiley, 2012.

Monbiot, G. *Regenesis: Feeding the World without Devouring the Planet.* London: Allen Lane, 1922.

Chapter 11

Grey, A. de. https://www.cambridgeindependent.co.uk/business/living-to-1-000-the-man-who-says-science-will-soon-defeat-ageing-9050845/

Sinclair, D. *Lifespan: Why We Age – and Why We Don't Have To.* New York: Atria Books, 2019.

Conclusion

Sinclair, D. *Lifespan: Why We Age – and Why We Don't Have To.* New York: Atria Books, 2019.

Sinclair, D., *Lifespan* book website: https://lifespanbook.com

Bibliography

Asprey, Dave. *The Bulletproof Diet: Lose Up to a Pound a Day, Reclaim Your Energy and Focus, and Upgrade Your Life.* Rodale Books, 2015.

Attia, Peter. *Outlive: The Science and Art of Longevity.* Vermilion, 2023.

Azhar, Azeem. *Exponential: Order and Chaos in an Age of Accelerating Technology.* Penguin, 2022.

Blackburn, Elizabeth and Epel, Elissa. *The Telomere Effect: A Revolutionary Approach to Living Younger, Healthier, Longer.* Orion Spring, 2018.

Blaser, Martin. *Missing Microbes: How Killing Bacteria Creates Modern Plagues.* Oneworld Publications, 2015.

Bostrom, Nick. *Superintelligence: Paths, Dangers, Strategies.* OUP Oxford, 2016.

Botsman, Rachel and Rogers, Roo. *What's Mine is Yours: How Collaborative Consumption is Changing the Way we Live.* Collins, 2011.

Campbell-McBride, Natasha. *Gut and Psychology Syndrome: Natural Treatment for Autism, Dyspraxia, A.D.D., Dyslexia, A.D.H.D., Depression, Schizophrenia.* Medinform Publishing, 2018.

Carey, Nessa. *Hacking the Code of Life: How Gene Editing Will Rewrite Our Futures.* Icon Books Ltd, 2020.

Chase, Callum. *The Economic Singularity: Artificial Intelligence and the Death of Capitalism.* Three Cs, 2016.

Chase, Callum. *Surviving AI: The Promise and Peril of Artificial Intelligence.* Three Cs, 2015.

Clear, James. *Atomic Habits: An Easy and Proven Way to Build Good Habits and Break Bad Ones.* Random House Business, 2018.

Collen, Alanna. *10% Human: How Your Body's Microbes Hold the Key to Health and Happiness.* William Collins, 2016.

Cooper, George. *Be Your Own Nutritionist: Rethink Your Relationship with Food.* Short Books Ltd, 2013.

D'Adamo, Peter and Whitney, Catherine. *Eat Right for Your Type.* Arrow, 2017.

David, Frank S. *The Pharmagellan Guide to Analyzing Biotech Clinical Trials.* Pharmagellan, 2022.

Davies, Dame Professor Sally. *The Drugs Don't Work: A Global Threat.* Penguin, 2013.

Dawkins, Richard. *The Blind Watchmaker.* Penguin, 2006.

Dawkins, Richard. *The Selfish Gene.* OUP Oxford, 2016.

Dawkins, Richard. *Unweaving the Rainbow: Science, Delusion and the Appetite for Wonder.* Penguin, 2006.

Diamandis, Peter & Kotler Steven. *Abundance: The Future Is Better Than You Think.* Free Press, 2014.

Diamandis, Peter & Kotler Steven. *Bold: How to Go Big, Create Wealth and Impact the World.* Simon & Schuster, 2016.

Diamandis, Peter & Kotler Steven. *The Future is Faster Than You Think: How Converging Technologies Are Transforming Business, Industries, and Our Lives.* Simon & Schuster, 2020.

Diamond, Jared. *Collapse: How Societies Choose to Fail or Survive.* Penguin, 2011.

Diamond, Jared. *Guns, Germs and Steel: A Short History of Everybody for the Last 13,000 Years.* Vintage, 1998.

Diamond, Jared. *The Rise and Fall of the Third Chimpanzee.* Vintage, 1992.

Doerr, John. *Speed & Scale: A Global Action Plan for Solving Our Climate Crisis Now.* Penguin, 2021.

Doudna, Jennifer and Sternberg, Samuel. *A Crack in Creation: The New Power to Control Evolution.* Vintage, 2018.

Enders, Giulia. *Gut: The Inside Story of Our Body's Most Under-Rated Organ.* Scribe UK, 2017.

Enriquez, Juan & Gullans, Steve. *Evolving Ourselves: Redesigning the Future of Humanity—One Gene at a Time.* Portfolio, 2016.

Ferguson, Niall. *The Ascent of Money: A Financial History of the World.* Penguin, 2019.

Ferriss, Tim. *The 4-Hour Body: An Uncommon Guide to Rapid Fat-Loss, Incredible Sex and Becoming Superhuman.* Vermillion, 2011.

Ferriss, Tim. *Tools of Titans: The Tactics, Routines, and Habits of Billionaires, Icons, and World-Class Performers.* Vermillion, 2016.

Ferriss, Tim. *Tribe of Mentors: Short Life Advice from the Best in the World.* Vermillion, 2017.

Gawande, Atul. *Being Mortal: Medicine and What Matters in the End.* Profile Books Ltd, 2015.

Gawande, Atul. *The Checklist Manifesto: How to Get Things Right*. Profile, 2011.

Goldacre, Ben. *Bad Pharma: How Medicine is Broken, and How We Can Fix It*. Fourth Estate, 2013.

Goldacre, Ben. *Bad Science*. Harper Perennial, 2009.

Goldacre, Ben. *I Think You'll Find It's a Little Bit More Complicated Than That*. Fourth Estate, 2014.

Greger, Michael & Stone, Gene. *How Not to Die: Discover the Foods Scientifically Proven to Prevent and Reverse Disease*. Pan, 2017.

Hands, John. *Cosmosapiens: Human Evolution from the Origin of the Universe*. Duckworth, 2016.

Hari, Johan. *Chasing the Scream. The Search for the Truth About Addiction*. Bloomsbury, 2019.

Hari, Johan. *Lost Connections: Why You're Depressed and How to Find Hope*. Bloomsbury, 2019.

Harari, Yuval. *Homo Deus; A Brief History of Tomorrow*. Vintage, 2017.

Harari, Yuval. *Sapiens: A Brief History of Humankind*. Vintage, 2015.

Harris, Dan. *10% Happier: How I Tamed the Voice in My Head, Reduced Stress Without Losing My Edge, and Found Self-Help That Actually Works - A True Story*. Yellow Kite, 2017.

Hof, Wim. *The Wim Hof Method: Activate Your Potential, Transcend Your Limits*. Rider, 2022.

Isaacson, Walter. *The Code Breaker: Jennifer Doudna, Gene Editing, and the Future of the Human Race*. Simon & Schuster UK, 2021.

Kahneman, Daniel. *Thinking, Fast and Slow*. Penguin, 2012.

Kelly, Kevin. *The Inevitable: Understanding the 12 Technological Forces That Will Shape Our Future*. Penguin, 2017.

Kotler, Steven. *The Rise of Superman: Decoding the Science of Ultimate Human Performance*. Quercus, 2015.

Kotler, Steven & Wheal Jamie. *Stealing Fire: How Silicon Valley, the Navy SEALs, and Maverick Scientists Are Revolutionizing the Way We Live and Work*. Harper Collins, 2018.

Kurzweil, Ray & Grossman, Terry. *Transcend: Nine Steps to Living Well Forever*. Rodale Press, 2019.

Littlehales, Nick. *Sleep: The Myth of 8 Hours, the Power of Naps … and the New Plan to Recharge Your Body and Mind*. Penguin, 2016.

Lomborg, Bjorn. *False Alarm: How Climate Change Panic Costs Us Trillions, Hurts the Poor, andFails to Fix the Planet*. Basic Books, 2021.

Mackay, David. *Sustainable Energy - Without the Hot Air.* Green Books, 2009.

Maddox, Brenda. *Rosalind Franklin: The Dark Lady of DNA*. Harper Collins, 2003.

Matten, Glen and Goggins, Aidan. *The Health Delusion: How to Achieve Exceptional Health in the 21st Century.* Hay House, 2012.

Mellon, Jim and Chalabi, Al. *Cracking the Code: Understand and Profit from the Biotech Revolution That Will Transform Our Lives and Generate Fortunes.* Wiley, 2012.

Mellon, Jim and Chalabi, Al. *Juvenescence: Investing in the Age of Longevity.* Fruitful Publications, 2017.

Mellon, Jim. *Moo's Law: An Investor's Guide to the New Agrarian Revolution.* Fruitful Publications, 2020.

Monbiot, George. *Regenesis: Feeding the World without Devouring the Planet.* Penguin, 2023.

Mosley, Michael. *Clever Guts Diet, the: How to revolutionise your body from the inside out.* Short Books, 2017.

Mukherjee, Siddhartha. *The Emperor of all Maladies: A Biography of Cancer.* Fourth Estate, 2011.

Nestor, James. *Breath: The New Science of a Lost Art.* Penguin, 2021.

Pinker, Steven. *Enlightenment Now: The Case for Reason, Science, Humanism and Progress.* Penguin, 2019.

Pollan, Michael. *Food Rules: An Eater's Manual.* Penguin, 2010.

Pollan, Michael. *How to Change Your Mind: The New Science of Psychedelics.* Penguin, 2019.

Pollan, Michael. *In Defence of Food. The Myth of Nutrition and the Pleasures of Eating.* Penguin, 2009.

Pollan, Michael. *The Omnivore's Dilemma: The Search for a Perfect Meal in a Fast-Food World.* Bloomsbury, 2011.

Ridley, Matt. *The Evolution of Everything: How Small Changes Transform Our World.* Harper Collins, 2016.

Ridley, Matt. *Genome: The Autobiography of a Species In 23 Chapters.* Fourth Estate, 2000.

Ridley, Matt. *How Innovation Works. Serendipity, Energy and the Saving of Time.* Fourth Estate, 2021.

Ridley, Matt. *Nature via Nurture: Genes, Experience and What Makes Us Human.* Harper Perennial, 2004.

Ridley, Matt. *The Rational Optimist: How Prosperity Evolves.* Fourth Estate, 2011.

Robson, David. *The Expectation Effect: How Your Mindset Can Transform Your Life.* Canongate Books, 2022.

Rosling, Hans and Ola. *Factfulness: Ten Reasons We're Wrong About the World - And Why Things Are Better Than You Think.* Sceptre, 2019.

Ross, Alec. *The Industries of the Future.* Simon & Schuster UK, 2017.

Schellenberger, Michael. *Apocalypse Never: Why Environmental Alarmism Hurts Us All.* Harper, 2020.

Sculley, John. *Moonshot!: Game-Changing Strategies to Build Billion-Dollar Businesses.* Rosetta Books, 2014.

Shapiro, Deb. *Your Body Speaks Your Mind: Understanding How Your Emotions and Thoughts Affect You Physically.* Piatkus Books, 2007.

Sinclair, David. *Lifespan: Why We Age and Why We Don't Have to.* Thorsons, 2019.

Starrett, Kelly and Juliet. *Built to Move: The 10 Essential Habits That Will Help You Live a Longer, Healthier Life.* Orion Spring, 2023.

Syed, Matthew. *Black Box Thinking: Marginal Gains and the Secrets of High Performance.* John Murray, 2016.

Pinker, Steven. *The Better Angels of our Nature: A History of Violence and Humanity.* Penguin, 2012.

Pinker, Steven. *Enlightenment Now. The Case for Reason, Science, Humanism, and Progress.* Penguin, 2019.

Thiel, Peter. *Zero to One: Notes on Start Ups, or How to Build the Future.* Virgin Books, 2015.

Thunberg, Greta. *No One is Too Small to Make a Difference.* Penguin, 2019.

Walker, Matthew. *Why We Sleep: The New Science of Sleep and Dreams.* Penguin, 2018.

Watson, James D. *The Double Helix.* W&N, 2010.

Werth, Barry. *The Billion Dollar Molecule: The Quest for the Perfect Drug.* Simon & Schuster, 1995.

Index

Acknowledgements

Getting this book from conception to some kind of finishing line has taken considerably more than two years. It has certainly been a labour of love. Those close to me know that it has been rather more 'labour' than 'love' than would have been ideal for much of that time.

As I'm sure is a fairly standard experience for many folk attempting to write a book, the process is an emotional roller-coaster to put it mildly. One day, sufficiently caffeinated and with suitably euphoric music blaring in your headphones, you 'flow' several thousand words of stuff you're really quite pleased about. The next thing you know, you're losing several hours of a Sunday morning stuck on one paragraph or even solitary sentence. A common unifying theme, however, is that you wouldn't be able to do any of it without the involvement, support and encouragement of a huge number of individuals.

First and foremost, my deepest gratitude goes to my wife, Rachel, who has lived through the highs and lows alongside me. I couldn't have got there without your unwavering support, understanding and tolerance of a pretty unusual and Spartan existence for the last two years. Thank you!

Thanks too to my nuclear family – Neville and Michael Craig, and Joanna (Banana) Nield - respectively my father, brother and sister. Thanks for always being there for your crazy son / brother.

I would also like to extend my heartfelt thanks to the teams at Plain English Finance and the Conviction Life Sciences Company: Tim Peacock, Roderick Collins, Dimitri Goulandris, Alan Back, Dave Graham, Geoff Miller, Dr Luke Zhou, Dr Victoria Gordon, John Whittle and Grant Cameron.

Thank you all for your belief and support and for tolerating my slight obsession with the future of the biotech industry. Thanks too to our compliance consultants Kim Ellis and Jon Wilson at Ellis Wilson Limited, and Andy Jay, Zihong Su, Rob Carson and the team at our accountants Fiander Tovell, given the crucial role you all play in what we do.

A special acknowledgment goes out to *all* of the shareholders of Plain English Finance, without whom this book could not have seen the light of day. My particular thanks to those of you who have been involved in the project in one way or another or whose friendship and support have kept me going along the way: Richard Allen, Iain Cullen, Andrew Stancliffe, David Curie, Steve Jackson, Spencer Crooks, Professors Stephen Thomas and Andrew Clare, Jeremy Smyth, Phil Webster, Matthew Kates, David Holdsworth, Ian Taylor and Kris Cudmore.

Thanks are due to the team at my publisher, Hodder & Stoughton/John Murray Press: Iain Campbell our publishing director for taking a chance on three books now, Meaghan Lim for so ably shepherding the production process and Robert Anderson and Antonia Maxwell for your vital editorial contributions.

I would also like to express my gratitude to the investment and other professionals and analysts who have so generously shared their time and knowledge over the years. Your valuable input has been instrumental in shaping so many elements of the book.

I'd like to make particular mention of Jim Birch, Rob Wiegold and Simon Niven at Shard Stockbrokers. You've always been such a delight to work with and deeply knowledgeable about the UK smaller companies scene. Similarly, Marcus Baker at Rothschild, Elliott Shaw and Andrew Keith at Shore Capital, Mike Seabrook, Josh Chandler and the team at Oberon, James (Woody) Wood at Winterflood, Sam Ogunsalu at Lancea Partners, Mike Dobson at Blue River, Damian Robinson at Cazenove, David Newlands at Newlands Capital, Freddy Crossley and Dr David Cox at Panmure Gordon, Franc Gregori at Trinity Delta, Nigel Birks at Cavendish, Joe Sluys and Siobhan McManus at SquareBook, Stuart Archibald at Jarden Partners in Sydney, Hayley Mullens and the team at h2Radnor, Emma Kane and team at SEC Newgate, plus Andrew Holder, Chris Stebbings, Virgil Wolf and Andrew McQuade.

Thanks too to my old team at WG Partners who were instrumental in my learning about the sector to begin with: David Wilson, Dr Nigel Barnes, Claes Spång, Chris Lee, Parthiv Patel, Jeff Glushakow and Olga Holme.

A heartfelt thank you to the management teams of so many companies I have met along the way. Your unstinting commitment to patient outcomes and to trying to build the future when faced with the never-ending challenges thrown at you by such an imperfect capital market environment have been genuinely inspiring. There are too many to name, but I'd like to make particular mention of: John Dawson, CBE, former CEO of Oxford Biomedica, Dr Alastair Smith and Dr Eliot Forster at Avacta, Andrew

Newland at Angle, Jon Pilcher at Neuren Pharma, Dr Sarah Howell and Susan Lowther at Arecor, Lisa Anson at RedX, Josh Fleet and the team at Soterios, Steve Harris and Darren Mercer at CS Pharma, Hugo Tewson at Digostics, Neil Clark, former CEO of Destiny Pharma, and Barry Smith and the team at Symprove.

I extend my appreciation and gratitude to the many inspiring authors whose works have fascinated, influenced and motivated me. I am enormously grateful to be able to stand on the shoulders of your collective wisdom. (Please see the bibliography for a full list).

...and finally to my little treasures, Ella May and Oscar William Craig. Your future is biotech!

ALSO BY ANDREW CRAIG

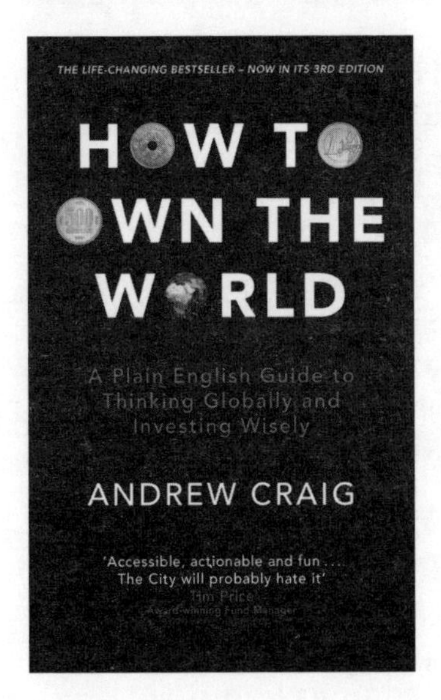

HOW TO OWN THE WORLD

THE LIFE-CHANGING PERSONAL FINANCE BESTSELLER THAT SHOWS YOU HOW TO MAKE MONEY FROM YOUR MONEY – NOW IN A REVISED 3RD EDITION.

Discover the money secret understood by virtually every rich person in history. Turn hundreds into millions through the power of compound interest.

HOW TO OWN THE WORLD shows you that:

- No one is better placed than you to make the most of your money.
- You can do better than many finance professionals.
- Making money from your money is easier than you think.
- You can make far more from your money than you ever thought possible.
- You can make more from your money than you can from your job.
- All this is possible no matter how much you currently earn.
- It's easier today than ever.
- It's time to start now.

It is entirely realistic for you to control your wealth, make a lot of money, and become financially free as a result. HOW TO OWN THE WORLD shows you how. With just a little knowledge you can turn your financial fortunes around and change your life.

'If you want just one book on investment from the cacophony, you couldn't do much better'
Michael Mainelli, Economics Professor, Lord Mayor of London

Trade Paperback ISBN 978-1-47369-5-306
ebook ISBN 978-1-47369-5-320
Audiobook ISBN 978-1-47369-5-337

Join Andrew's Free Email List?

If you enjoyed this book and would like to hear more from Andrew, please do consider subscribing to his free email list.
You just need to enter your name and email address from the home page of his website: www.plainenglishfinance.com

Would you like your people to read this book?

If you would like to discuss how you could bring these ideas to your team, we would love to hear from you. Our titles are available at competitive discounts when purchased in bulk across both physical and digital formats. We can offer bespoke editions featuring corporate logos, customized covers, or letters from company directors in the front matter can also be created in line with your special requirements.

We work closely with leading experts and organizations to bring forward-thinking ideas to a global audience. Our books are designed to help you be more successful in work and life.

For further information, or to request a catalogue, please contact: **business@johnmurrays.co.uk**

John Murray Business is an imprint of John Murray Press.